CLUSTER ANALYSIS AND DATA MINING

CLUSTER ANALYSIS AND DATA MINING
An Introduction

R.S. King

MERCURY LEARNING AND INFORMATION
Dulles, Virginia
Boston, Massachusetts
New Delhi

Publisher: David Pallai
MERCURY LEARNING AND INFORMATION
22841 Quicksilver Drive
Dulles, VA 20166
info@merclearning.com
www.merclearning.com
1-800-758-3756

This book is printed on acid-free paper.

R.S. King. *Cluster Analysis and Data Mining: An Introduction*
ISBN: 978-1-938549-38-0

Library of Congress Control Number: 2014941165

141516321   Printed in the United States of America

Our titles are available for adoption, license, or bulk purchase by institutions, corporations, etc. Digital versions of this title are available at www.authorcloudware.com. Companion disc files may be obtained by contacting info@merclearning.com. For additional information, please contact the Customer Service Dept. at 1-800-758-3756 (toll free).

To the memory of my father, who made it possible and to LaJuan, the shining light in my life who survived the process.

CONTENTS

PREFACE

This book is appropriate for a first course in clustering methods and data mining. Clustering and data mining methods are applicable in many fields of study, for example:

1. in the life sciences for developing complete taxonomies,

2. in the medical sciences for discovering more effective and economical means for making positive diagnosis in the treatment of patients,

3. in the behavioral and social sciences for discerning human judgments and behavior patterns,

4. in the earth sciences for identifying and classifying geographical regions,

5. in the engineering sciences for pattern recognition and artificial intelligence applications, and

6. in decision and information sciences for analysis of markets and documents.

The first five chapters consider early historical clustering methods. Chapters 1 and 2 are an introduction to general concepts in clustering methods, with an emphasis on proximity measures and data mining. Classical numerical clustering methods are presented in Chapters 3 and 4: hierarchical and partitioned clustering. These methods are particularly defined only on numeric data files. A clustering method implemented via multiple linear regression, judgmental analysis (JAN), is discussed in Chapter 5. JAN allows for numerical and categorical variables to be included in a clustering study.

All of the methods in Chapters 1 through 5 generate partitions on a study's data file, referred to as *crisp clustering* results. *Fuzzy clustering* methods presented in Chapter 6, capture partitions plus modified versions for the partitions. The modified partitions allow for overlapping clusters.

Chapter 7 is an introduction to the data mining topics of classification and association rules, which enable qualitative rather than simply quantitative data mining studies to be conducted.

Cluster analysis is essentially an art, but can be accomplished scientifically if the results of a clustering study can be validated. This is discussed in Chapter 8. Determination of the validity of individual clusters and the validation of a clustering, or collection of clusters, are discussed.

Chapter 9 surveys a variety of algorithms for clustering categorical data: ROCK, STIRR, CACTUS, and CLICK. These methods are dependent on underlying data structures and are applicable to relational databases.

Applications of clustering methods are presented in Chapters 10 through 11. Chapter ten discusses classical statistical methods for identifying outliers. Additionally, crisp and fuzzy clustering methods are applied to the outlier identification problem. Chapter 11 is an overview of model-based clustering. This is often used in physical science research studies for data generation.

A summary of the issues and trends in the cluster analysis field is made in Chapter 12. Besides giving recommendations for further study, an introduction to neural networks is presented. The appendices provide a variety of resources (software, URLs, algorithms, references) for the cluster analysis plus URLs for test data files.

The text is applicable to either a course on clustering, data mining, and classification or as a companion text for a first class in applied statistics. Clustering and data mining are good motivators and applications of the topics commonly included in an introductory applied statistics course.

The scheduling references for each of the chapters, in an applied statistics class, could be as follows:

Chapters 1-4: after study of descriptive statistics.

Chapter 9: immediately following Chapters 1-3.

Chapter 6: after study of descriptive statistics.

Chapter 10: after studying the Empirical Rule and Chebychev's Law.

Chapter 7: after studying probability.

Chapter 8: after study of hypothesis testing.

Chapter 5: after study of correlation, and both linear and multiple linear regression.

Chapter 11: after study of statistical inference.

No previous experience or background in clustering is assumed. Elementary statistics plus a brief exposure to data structures are the prerequisites. Informal algorithms for clustering data and interpreting results are emphasized. In order to evaluate the results of clustering and to explore data, graphical methods and data structures are used for representing data. Throughout the text, examples and references are provided, in order to enable the material to be comprehensible for a diverse audience.

INTRODUCTION TO CLUSTER ANALYSIS

1.1 WHAT IS A CLUSTER?

Many of the decisions being made today involve more than one person. An important question in the group decision process is: "How does the group arrive at its final decision?" There have been a number of different mathematical and statistical approaches used by researchers attempting to model the decision-making process including game theory, information theory, and linear programming. Due to the large variety of decision-making situations, different types of decision processes, and the kinds of skills required, there is still a great deal of concern about the best way to make decisions. In many cases there is no objective approach. The individuals in the decision-making group each use their own set of criterion in reaching a decision. This approach might work in a situation where a consensus is not needed. However, in the case where a single group decision is needed, there must be a "meeting of the minds."

One approach used is the *Delphi Technique*. This technique was designed in the early 1950s by the Rand Corporation to predict future outcomes. It is a group information gathering process to develop consensus opinion from a panel of experts on a topic of interest. In the normal Delphi scenario, the panel never meets face to face but interacts through questionnaires and feedback. This noncontact approach alleviates the worry over such issues as individual defensiveness or persuasiveness. However, opinions can be swayed due to a participant observing the responses of the rest of the panel. Another problem with the Delphi Technique is that the noncontact aspect is not feasible when, for example, the panel is the graduate admissions committee at a university.

Cluster analysis is another technique that has been used with success in the decision-making process. First, the investigator must determine the answer to "What is a cluster?" The **premise in cluster analysis** is: given a number of individuals, each of which is described by a set of numerical measures, devise a classification scheme for grouping the objects into a number of classes such that the objects within classes are *similar* in some respect and *unlike* those from other classes. These deduced classes are the clusters. The number of classes and the characteristics of each class must be determined from the data as discussed by Everett.[1]

The key difference between cluster analysis and the Delphi Technique is that cluster analysis is strictly an objective technique. Whereas individual decisions can be swayed in an attempt to reach consensus in the Delphi process, or a "happy medium" is reached which does not really portray the feelings of the group as a whole. This is not the case in cluster analysis. Clusters of individuals are reached using an objective mathematical function. One particular type of cluster analysis called Judgmental ANalysis (JAN) takes the process one step further. Not only does it classify the panel into similar groups based on a related regression equation, but it also allows for these equations to be combined into a single policy equation. The JAN technique has been in use since the 1960s. It has proven to be an effective first step for methods of capturing and clustering the policies of judges.

Attempts at classification, that is sorting similar things into categories, can be traced back to primitive humans. The ability to classify is a necessary prerequisite for the development of language. Nouns, for example, are labels

---

[1] Everitt, B. S. (1980). *Cluster analysis (2nd ed.)*. New York: Halsted Press.

used to classify a particular group of objects. Saying that a particular four-legged animal is a "dog" allows us to put that animal into a category separate from cats, sheep, and horses. In other words, it allows us to communicate.

The classification of people and animals is almost as old as language. The early Hindus categorized humans into six types based on sex, physical, and behavioral characteristics. The early Greeks and Romans used classification to get a better understanding of the world around them. Galen, A.D. 129-199, defined nine temperamental types that were assumed to be related to a person's susceptibility to various diseases and to individual differences in behavior as discussed by Everitt.[1] Development of a method to categorize animals into species was initiated by Aristotle. He started by dividing them into red blooded (vertebrates) and those not having red blood (invertebrates). He then subdivided the two groups again based on how their young were born. Theophrastus continued Aristotle's work, providing the groundwork for biological research for centuries. Eventually, new taxonomic systems were developed by such people as Linnaeus, Lindley, and Darwin. Classification was not restricted to the biological sciences. In chemistry, Mendeleyev used classification to develop the periodic tables, discussion by Everitt.[1]

In the 1960s, two events led to an explosion of interest in cluster analysis. The availability and spread of large, high-speed computers opened up new possibilities for researchers. Additionally, the publication of *Principles of Numerical Taxonomy* by Sokal and Sneath[2] covered the following three important areas:

1. a number of different cluster analysis techniques

2. the use of computers in classification research

3. a radically empirical approach to biological taxonomy presented by Blashfield and Aldenderfer[3]

The need for cluster analysis arises in many fields of study. For example, Anderberg[4] lists six areas where cluster analysis has been used successfully:

1. In the life sciences (biology, botany, zoology, etc.), the objects of analysis are life forms such as plants, animals, and insects. The

[2] Sokal, R. R., and Sneath, P. H. A. (1963). *Principles of Numerical Taxonomy*. W. H. Freeman.

[3] Blashfield and Aldenderfer, M. S. (1978). The literature on cluster analysis. *Multivariate Behavioral Research*, 13, 271-295.

[4] Anderberg, M. R. (1973). *Cluster analysis for applications*. New York: Academic Press.

operational purpose of the analysis may range from developing complete taxonomies to delimiting the subspecies of a distinct but varied species.

2. In the medical sciences (psychiatry, pathology, etc.), the objects of a cluster analysis may be diseases, patients, symptoms, and laboratory tests. The operational emphasis here is on discovering more effective and economical means for making positive diagnosis in the treatment of patients.

3. In the behavioral and social sciences (psychology, sociology, education, etc.), some of the wide variety of objects of analysis are training methods, behavior patterns, organizations, human judgments, families, and teaching techniques.

4. Applications of cluster analysis in the earth sciences (geology, geography, etc.) have included the study of land and rock formations, soils, river systems, cities, and regions of the world.

5. Examples of entities that have been clustered in the engineering sciences (pattern recognition, artificial intelligence, cybernetics, electrical engineering, etc.) include handwritten characters, speech, fingerprints, electrocardiograms, radar signals, and circuit designs.

6. In the area of information and decision sciences (information retrieval, political science, economics, marketing research, operational research, etc.), cluster analysis has been applied to the analysis of documents, markets, investments, and credit risks.

As can be seen, the areas in which cluster analysis has been used with success are large and varied. It is also interesting to note some of the other names for cluster analysis in these different fields. Some of the aliases mentioned by Anderberg[3] are numerical taxonomy (biology, botany, ecology), typology (social sciences), learning without a leader (pattern recognition, cybernetics, electrical engineering), clumping (information retrieval, linguistics), regionalization (geography), partition (graph theory, circuit designers), and serration (anthropology). The reasons for clustering are as many and as varied as the fields and names. Everitt[1] mentions seven possible uses of clustering techniques including data reduction, data exploration, hypothesis generating, hypothesis testing, model fitting, and prediction based on groups, and finding a true typology.

1.2 CAPTURING THE CLUSTERS

Cluster analysis employs a measure of similarity or dissimilarity for assigning points in space to a cluster. In general terms, points exist in a space (which could be a plane, the surface of a sphere, three-dimensional space, etc.) that relate to the concept of distance that matches geometrical intuition. Formally, the operational definition of this type of distance is: a **proximity measure** in space $M = \{A, d\}$ consists of a nonempty set A together with a distance function $d: A \times A \rightarrow \mathbf{R}^2$ which satisfies:

1. $d(x, y) \geq 0; d(x, y) = 0$ if and only if $x = y$

That is, the distance between two distinct points is strictly positive.

2. $d(x, y) = d(y, x)$ for all x, y in A

The distance from x to y is equal to the distance from y to x.

3. (a) for a dissimilarity $d(i,i) = 0$, for all i

The distance between a point and itself is zero, or

points aren't different from themselves.

(b) for a similarity $d(i,i) \geq max_k\, d(i, k)$ for all i

The points are most similar to themselves.

So, what does this actually mean? First, there must exist a nonempty set A, basically a collection of one or more points. Given a distance function, d, which can be used to determine the distance between any two points of A, d, must also follow certain rules.

The first rule states that one cannot have a negative distance and the distance between two points can only be zero if the two points are, in fact, in exactly the same place. The second rule states that the distance between two points must be the same for whichever direction is measured, going from x to y covers the same distance as going from y to x. Finally, measurement is either based upon similarity or dissimilarity between points.

Let $M = \{A, d\}$ be the space being studied, let a be in A, and let $\varepsilon > 0$. The ε-**neighborhood** of a in M is defined to be:

$$\mathbf{N}_\varepsilon(a) = \{x \text{ in } A \mid d(x, a) < \varepsilon\}$$

That is, the collection of points x in A within distance ε of a.

It is worth emphasizing that $N_\varepsilon(a)$ does *not* include the boundary. It consists *only* of the interior of the "neighborhood." If the boundary is included, the neighborhood is called a *closed* neighborhood.

In $\mathbf{R}^2$ (the plane), $N_\varepsilon(a)$ is the interior of a disc of radius ε centered on a.

In $\mathbf{R}^3$ (three-dimensional space), $N_\varepsilon(a)$ is the interior of a solid ball of radius ε centered on a.

The previous two examples have all used the Euclidean metric, that is, our intuitive notion of distance:

$$d_1 = d(i, k) = \left(\sum_{j=1}^{d} \left| x_{ij} - x_{kj} \right|^2 \right)^{1 \backslash 2}$$, in two-dimensional space and d is the number of features.

In $\mathbf{R}^2$, using the **Minkowski measure**,

$$d_2 = d(i, k) = \left(\sum_{j=1}^{d} \left| x_{ij} - x_{kj} \right|^r \right)^{1 \backslash r}$$, where d is the number of features, n is the number of patterns, and $r = 2$ is the dimension of the space, $N_\varepsilon(a)$ is the interior of a square centered on a, with sides of length 2ε parallel to the co-ordinate axes.

The **"sup" distance,**

$$d_3 = d(i, k) = max_{i \le j \le d} \left| x_{ij} - x_{kj} \right|$$, where d is the number of patterns, generates diamond shaped ε-neighborhoods.

Clusters are captured by attempting to find nonoverlapping ε-neighborhoods using the given proximity. The goal is to group objects in a group (or related) to one another and different from (or unrelated to) the objects in other groups. The greater the similarity (or homogeneity) within a group, and the greater the difference between groups, the finer granularity is present in the clustering.

Other commonly used proximity measures include:

City-block (Manhattan) distance. This distance is simply the average difference across dimensions. In most cases, this distance measure yields results similar to the simple Euclidean distance. The city-block distance is: distance $(x,y) = \sum_i \left| y_i - x_i \right|$

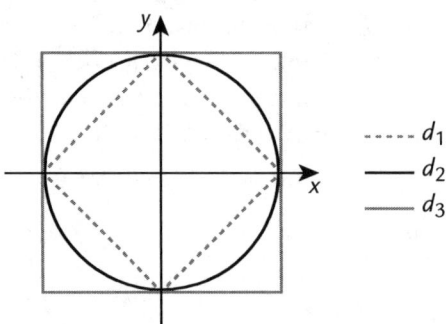

- - - - - d_1
————— d_2
————— d_3

FIGURE 1.1 Example ε-Neighborhoods.

Chebychev distance. This distance measure may be appropriate in cases when defining two objects as "different" if they are different on any one of the dimensions. The Chebychev distance is computed as: distance $(x,y) = \text{Maximum } |x_i - y_i|$

Power distance. Sometimes the emphasis is to increase or decrease the progressive weight that is placed on dimensions for different objects. This can be accomplished via the *power distance*. The power distance is computed as: distance $(x,y) = \sum_i \left(|x_i - y_i|^p \right)^{1/r}$, where r and p are user-defined parameters. Parameter p controls the progressive weight that is placed on differences on individual dimensions, parameter r controls the progressive weight that is placed on larger differences between objects. If r and p are equal to two, then this distance is equal to the Euclidean distance.

Percent disagreement. This measure is particularly useful if the data for the dimensions included in the analysis are categorical in nature. This distance is computed as: distance $(x,y) = (\text{Number of } x_i \neq y_i)/i$

Consider the following set of points in two-dimensional Euclidean space:

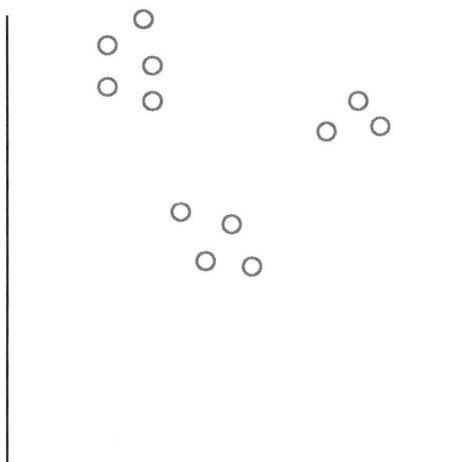

FIGURE 1.2 Points in Two-dimensional Euclidian Space.

Three groups would be identified with Euclidian neighborhoods.

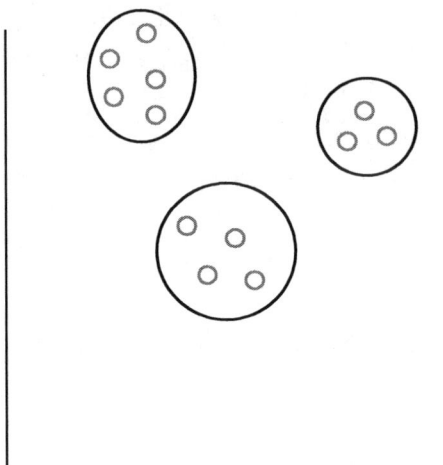

FIGURE 1.3 Three Groups Captured by Euclidean Neighborhoods.

Several primary questions need to be investigated when capturing the clusters. These questions include:

1. "How many clusters are present?" Consider the following situation:

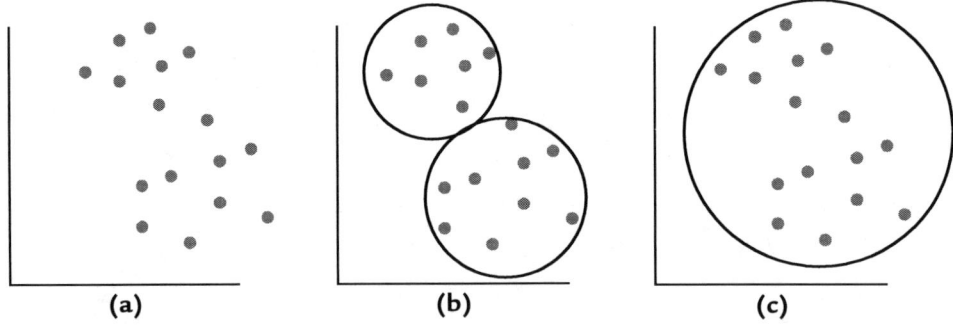

FIGURE 1.4 (a) Original Points, (b) Two Groups, and (c) One Group.

2. "Does the current ε-neighborhood and proximity measure correctly identify the clusters?"

For instance, only one cluster can be captured for the following set of points in two-dimensional Euclidean space:

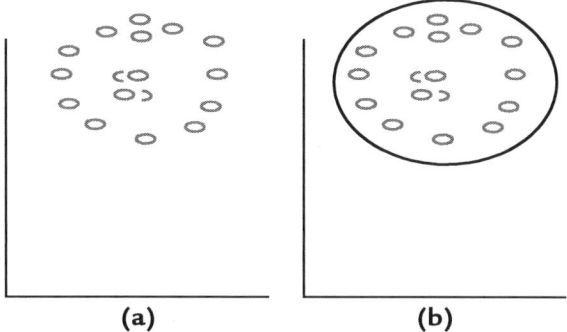

FIGURE 1.5 (a) Original Points and (b) One Cluster Interpretation.

In reality, the actual number of clusters should be two. In this case, the Euclidean neighborhoods are incapable of obtaining the correct number of clusters.

This example illustrates that cluster analysis is sensitive

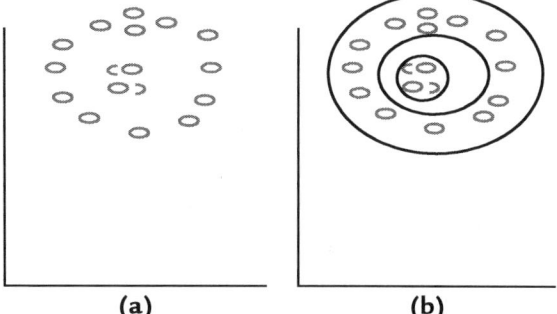

FIGURE 1.6 (a) Original Points and (b) Actual Groups.

to both the proximity measure selected and related ε-neighborhood shapes. Different approaches may yield different results. Consequently, the distance metric should be chosen carefully. The results should also be compared to analyses based on different proximity measures to enable determination of the robustness of the results.

3. Do the captured clusters have realistic interpretations?

4. Do any of the clusters overlap? If so, to what degree?

1.3 NEED FOR VISUALIZING DATA

The discussion on proximity measures for capturing clusters demonstrates that clustering software needs features that make clustering practical for a wide variety of applications. Such a package should at least provide highly optimized implementations of agglomerative, k-means, and graph clustering, especially in the context of sparse high-dimensional data. Additionally, the package should help the user sort through the algorithm options and resulting data files by providing an intuitive graphical interface. Clustering software should provide both standard statistics and unique visualizations for interpreting clustering results. Given the wide range of options and factors that are involved in clustering, the user should carefully analyze his results and compare them with results generated with different options. Visualizations enhance the analysis and comparisons.

1.4 THE PROXIMITY MATRIX

According to *Oxford Dictionary of Statistics*, a square matrix in which the entry in cell (j, k) is some measure of the similarity (or distance) between the items to which row j and column k correspond. A simple example would be a standard mileage chart—the smaller the entry, the closer together are the two items. Proximity matrices form the data for multidimensional scaling. Asymmetric matrices can occur (for example, if the measurement is time taken, then the journey from top to bottom of a hill will be shorter than the journey from bottom to top).

Suppose we are given the following **ordinal proximity matrix**:

	x_1	x_2	x_3	x_4	x_5
x_1	0	6	8	2	7
x_2	6	0	1	5	3
x_3	8	1	0	10	8
x_4	2	5	10	0	4
x_5	7	3	9	4	0

TABLE 1.1 An Ordinal Proximity Matrix.

The objects x_i, for the cell the value represents the Euclidean distance between the objects. Next we construct a **proximity ratio matrix**; a matrix where a proximity measure has been derived from the proximity matrix. The Euclidean proximity measure would generate the following proximity ratio matrix for the matrix in Table 1.2:

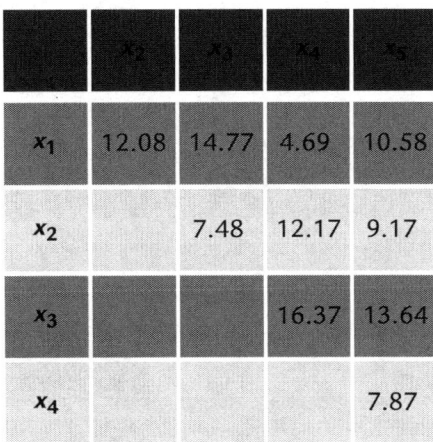

	x_2	x_3	x_4	x_5
x_1	12.08	14.77	4.69	10.58
x_2		7.48	12.17	9.17
x_3			16.37	13.64
x_4				7.87

TABLE 1.2 A Proximity Ratio Matrix.

Using the proximity ratio matrix we can perform a hierarchical clustering $\gamma_0, \gamma_2, \dots, \gamma_{n-1}$ where the mth clustering contains $n - m$ clusters. A level function, records the proximity for each clustering formed. For the start of this process, $L(k) = k$, because the levels are evenly spread apart.

$$L(m) = \min\{d(x_i, x_j) \mid \gamma_m \text{ is defined}\}$$

where the mth clustering contains $n - m$ clusters:

$$\gamma_{Cm} = \left\{\gamma_{m1}, \gamma_{m2}, \dots, \gamma_{m(n-m)}\right\}$$

The **cophenetic proximity measure** d_c on the n objects is the level at which objects x_i and x_j are first in the cluster.

$$d_c(i,j) = L(k_{ij})$$

where

For single-link, use $k_{ij} = min\left\{m : \left(x_i, x_j\right) \varepsilon \gamma_{Cmq} \text{ for some } q\right\}$

For complete-link, use $k_{ij} = max\left\{m : \left(x_i, x_j\right) \varepsilon \gamma_{Cmq} \text{ for some } q\right\}$

This process generates the following results for the complete-link solution:

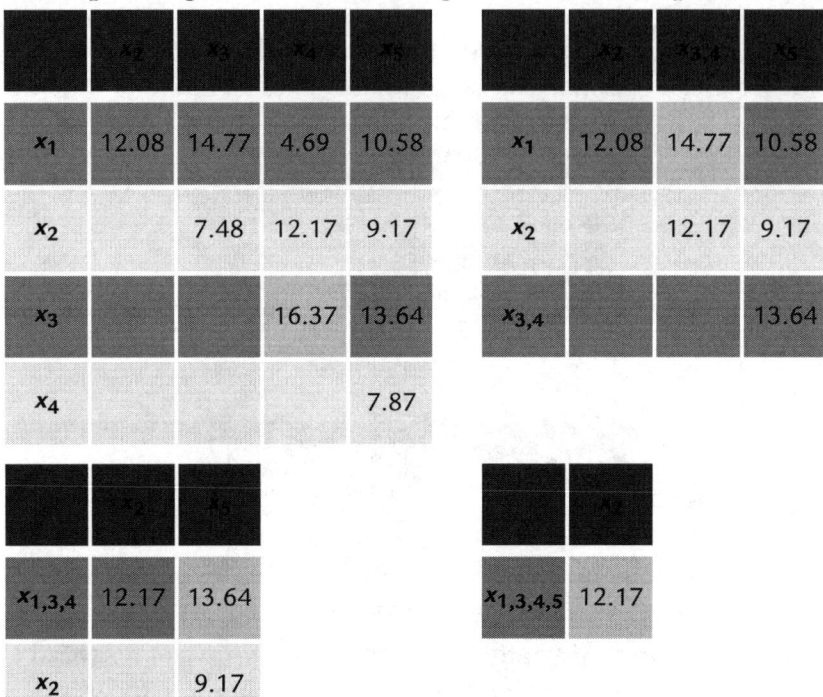

	x_2	x_3	x_4	x_5
x_1	12.08	14.77	4.69	10.58
x_2		7.48	12.17	9.17
x_3			16.37	13.64
x_4				7.87

	x_2	$x_{3,4}$	x_5
x_1	12.08	14.77	10.58
x_2		12.17	9.17
$x_{3,4}$			13.64

	x_2	x_5
$x_{1,3,4}$	12.17	13.64
x_2		9.17

	x_2
$x_{1,3,4,5}$	12.17

FIGURE 1.7 Complete-Link Clustering Process.

1.5 DENDROGRAMS

A special type of tree structure, called a **dendrogram**, provides a graphical presentation for the clustering. A dendrogram consists of layers of nodes, where each node represents a cluster. Lines connect nodes representing clusters which are nested together. Horizontal slices of a dendrogram indicate a clustering. For the latter complete-link clustering, we have:

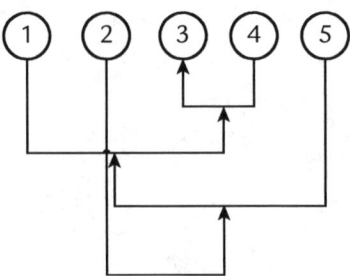

FIGURE 1.8 Dendrogram for a Complete-Link Clustering.

At the start $L(1) = 16.37$ for the clustering $\{(x_1), (x_2), (x_3), (x_4), (x_5)\}$.

On the first iteration, $L(2) = 14.77$ for the clustering $\{(x_1), (x_2), (x_3, x_4), (x_5)\}$.

After completion of the second iteration, $L(3) = 13.64$ for the clustering $\{(x_2), (x_1, x_3, x_4), (x_5)\}$.

Completion of the third iteration generates: $L(4) = 12.17$ for the clustering $\{(x_2), (x_1, x_3, x_4, x_5)\}$.

The last iteration generates: $L(5) = 12.17$ for the clustering $\{(x_1, x_2, x_3, x_4, x_5)\}$.

Clearly several obvious questions arise at this point in the cluster analysis:

- How many groups are present in the data?

- What is the group membership interpretation?

- Will different grouping algorithms have a common clustering result? Not necessarily!

- CLUSTERING IS AN ART AS WELL AS A SCIENCE!

1.6 SUMMARY

Cluster analysis is the formal study of algorithms and methods for grouping, or classifying objects.

Questions to resolve include:

What defines similarity between objects or between clusters?

What defines a distance between two clusters?

How can you capture clusters? What are the different methods for identifying these clusters?

When is it "best" to partition or identify clusters?

When is it "best" to stop joining clusters?

What are the right data elements to utilize in clustering for a problem in a specific application domain where we may have hundreds of variables?

What are the limitations of cluster analysis?

Basically the steps completed in clustering include:

Step One: Form similarities between all pairs of the objects based on a given attribute.

Step Two: Groups are constructed where within-group similarities are larger than the between-group similarities.

Let $d(i,k)$ be a proximity between the ith and kth patterns. Then:

- For a dissimilarity $d(i,i) = 0$, for all i.

- For a similarity, $d(i,i) = \max d(i,k)$, for all (i,k).

- $d(i,k) = d(k,i)$, for all (i,k).

- $d(i,k) \geq 0$, for all (i,k).

The common proximity measures are:

■ Minkowski metric: $d(i, k) = \left(\sum_{j=1}^{d} \left| x_{ij} - x_{kj} \right| \right)$

■ Euclidean distance: $d(i, k) = \left(\sum_{j=1}^{d} \left(x_{ij} - x_{kj} \right)^2 \right)^{1/2}$

■ Manhattan distance: $d(i, k) = \sum_{j=1}^{d} \left| x_{ij} - x_{kj} \right|$

■ Sup distance: $d(i, k) = \max_{1 \leq j \leq d} \left| x_{ij} - x_{kj} \right|$

1.7 EXERCISES

1. Suppose you are given two decks of playing cards, one with a blue backing and the other with a red backing. Discuss ways in which the 104 cards without jokers or 108 cards with jokers can be clustered. Is it possible to form a clustering which is not a partition?

2. What is the distinction between the following terms: similarity measure, dissimilarity measure, metric, and distance measure?

3. Complete the computations for the single-link clustering presented in the chapter.

4. Complete a single-link cluster on the ordinal proximity matrix given in Table 1.2.

5. Perform a complete-link cluster for each proximity measure given the following example considering the following data {(16,19), (20,23), (8,20), (1,23), (18,6), (5,28)} with associated labels {1, 2, 3, 4, 5, 6}, which are points within two-dimensional Euclidean space:

(a) Minkowski metric   (b) Manhattan distance

(c) Sup distance   (d) Percent disagreement

6. The cluster analysis presented in this chapter is a bottom-to-top (agglomerative) process. Perform a top-to-bottom (divisive) process. In other words, start with one cluster containing all the objects and finish with a clustering containing all the singleton clusters.

7. For problems 3 through 5, discuss how to determine when the clustering stops.

8. How could clustering methods be used for identifying outlier(s)? Note that outlier(s) by itself (themselves) will be a cluster. Think of an example of a tree diagram which will point out few outliers; and how the grouping pattern and the stem will be represented by a cluster of outliers.

9. What is the relationship between the linkage distance measure and the number of clusters?

10. Why would the number of clusters not be a simple continuously increasing number? Is it possible that there may not be a one-to-one relationship between the linkage distance and the number of clusters?

11. How does variable selection play a role in cluster analysis; what method is best to use?

12. Why is linkage distance inversely related to the number of clusters in general?

13. What happens if a similarity measure is used instead of distance measure?

14. What is meant by similarity or distance measures when we have qualitative data?

15. What is the major problem with the nonhierarchical method? (Hint: start point of the seed or center of the cluster)

16. Why should you standardize the data when doing cluster analysis?

17. Discuss how to use the dendrogram. (Tree structure for identifying the distance "between clusters" and which observations belong to which cluster—a graphical representation issue.)

18. Various factors affect the stability of a clustering solution, including: selection, distance/similarity measure used, different significance levels, and type of method (divisive vs. agglomerative) among others. Do some background research and present a report on a method that converges to the right solution in the midst of the above mentioned parameters.

19. You are given the following contingency table for binary data:

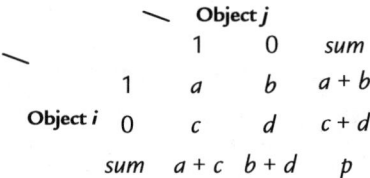

TABLE 1.3 Contingency Table.

    (a) Define a symmetry measure based upon the contingency table.

    (b) Define an asymmetry measure based upon the contingency table.

20. Describe the type of variables in the following table and define a dissimilarity measure for the binary variables.

Name	Gender	Fever	Cough	Test-1	Test-2	Test-3	Test-4
Jack	M	Y	N	P	N	N	N
Mary	F	Y	N	P	N	P	N
Jim	M	Y	P	N	N	N	N

21. Define a similarity measure for nominal variables.

22. What is an ordinal variable? Discuss how a dissimilarity measure can be defined for an ordinal variable.

23. What is a ratio-scaled variable? Discuss how nonlinear scale, such as an exponential scale, should be represented in a cluster analysis.

24. Databases consist of variables of mixed types. Discuss how to operationally define similarity/dissimilarity measures for a typical database.

25. What is a vector object? How can one operationally define a dissimilarity measure for vector objects?

CHAPTER 2

OVERVIEW OF DATA MINING

In This Chapter

2.1 WHAT IS DATA MINING?

A primary goal for many twenty-first century companies is to simultaneously maximize their rate of return and customer satisfaction. Supply chain management coupled with the associated entity relationship program enable firms to be competitive in the workforce. These firms are able to deliver a high-quality product that is highly useful to the customer in a timely fashion. Success is based upon understanding their customers, vendors, and supply chain.

Often this type of understanding is partially obtained by drilling-into-the-database. For example, consider a database for a chain of grocery stores. The top ten customers could be found simply by using filters. Pivot table processing would enable the identification of factors for purchasing habits broken down by geographical region and specific product lines. Graphical software would allow for the results to be visually displayed. These types of tasks, discussed in this paragraph, are in the realm of exploratory data analysis.

Today, data mining refers to the extraction of mathematical patterns from large databases. Methods from the fields of computer science, statistics, and machine learning comprise the data mining toolset. Rule sets, associations, sequenced associations, correlations, trends, and prediction models are some of the extracted mathematical patterns. Essentially data mining tools enable finding patterns in data and possibly inferring rules based upon the patterns.

SQL-based queries, use of human judgment, and online analytic processing are the simplest data mining tools. Classification via decision trees, cluster analysis, and regression comprise the next level of data mining tools. Another level of data mining tools includes neural networks and fuzzy systems.

A brief list of data mining methods includes: classification, clustering, associations, sequence discovery, regression, and forecasting. The objective of classification is to analyze the historical data stored in the database and generate a model for predicting future behavior. The objective of clustering is to partition a database into segments where members share similar features. Associations establish relationships that occur together in a given transaction. Sequence discovery's objective is to determine associations that occur over time. Regression and forecasting are useful for prediction.

2.2 DATA MINING RELATIONSHIP TO KNOWLEDGE DISCOVERY IN DATABASES

To successfully mine for knowledge in large databases, a framework or process should be followed. Knowledge Discovery in Databases (KDD) is a machine learning process that performs rule induction on a related procedure to establish knowledge from large databases. Fayyad, Piatesky-Shapiro, and Smyth[1] defined KDD as a process using data mining methods to find useful information and patterns in data. On the other hand, data

---

[1] Fayyad, U., Piatesky-Shaapiro, G., and Smyth, P. (1996). The KDD process for extracting useful knowledge from volumes of data. *Communications of the ACM*, 39(11):27-34.

mining uses algorithms to identify patterns in data derived through the KDD process. Organizational data is input to the KDD process in the enterprise data warehouse. Similar to structured programming, there exists a single source for data to be mined.

Fayyad, Piatesky-Shapiro, and Smyth[1] developed the following nine step model, which is iterative with many loops between any two steps:

Step 1: Fully ***develop and understand the application*** by learning relevant knowledge and related end-user goals.

Step 2: ***Select a target data set*** based on the features and actual data points to serve as the discovery data set.

Step 3: ***Perform data cleaning and preprocessing,*** which includes the tasks of removing outliers and noise, as well as dealing with missing values.

Step 4: By application of ***data reduction and projection*** methods obtain an invariant representation of the data.

Step 5: ***Select the data mining method*** that meets Step 1 requirements.

Step 6: ***Select the data mining algorithm*** based upon appropriate models and parameters of the chosen methods.

Step 7: ***Perform the data mining.***

Step 8: ***Interpret the mined patterns***.

Step 9: ***Incorporating the discovered knowledge into the performance system and documenting and reporting it to the stakeholders***. Note that the interpreted patterns must be checked for and resolved with respect to potential conflicts for previously believed knowledge.

Cabena, Hadjinian, Stadler, Verhees, and Zanasi[2] discuss an industrial model called CRISP-DM (Cross-Industry Standard Process for Data Mining), which was established by four companies: Integral Solutions Ltd.,

---

[2] Cabena, P., Hadjinian, P., Stadler, R., Verhees, J., and Zanasi, A. (1998). *Discovering Data Mining From Concepts to Implementation*. Prentice Hall, Saddle River, New Jersey.

NCR, DaimlerChrysler, and OHRA. This KDD process incorporates a description of business aspects as well as the technical description.

The CRISP-DM KDD model consists of the following steps:

Step 1: Develop a business understanding:

1. Determine the business objectives.

2. Make an assessment of the problem.

3. Determine the data mining goals.

4. Generate a project plan.

Step 2: Understand the data:

1. Collect the initial data.

2. Completely describe the data.

3. Perform preliminary exploration of the data.

4. Perform data quality verification.

Step 3: Data preparation:

1. Select the project data.

2. Perform data cleaning.

3. Construct the new project attributes.

4. Transform the data if needed.

Step 4: Modeling:

1. Selection of modeling techniques.

2. Development of test design.

3. Assessment of selected modeling techniques.

Step 5: Evaluation:

1. Evaluate the results.

2. Perform a process review.

3. Determine the next step to be taken.

Step 6: Deployment:

> **1.** Develop a deployment plan.
>
> **2.** Develop a monitoring and maintenance plan.
>
> **3.** Generate the final report.
>
> **4.** Review the process substeps.

2.3 THE DATA MINING PROCESS

In summary, data mining is the process of extracting valid, authentic, and relevant patterns from large databases. The basic steps in data mining are:

1. Problem definition.

2. Data preparation.

3. Data exploration.

4. Model selection and/or construction.

5. Model exploration and validation.

6. Model deploying and updating.

2.4 DATABASES AND DATA WAREHOUSING

The data miner needs an integrated company-wide view of high-quality data and information. Informational systems need to be separated from operational systems in order to improve corporate data management. Data warehousing is one method organizations use to integrate data to gain greater data accessibility across the organization.

Definition: A *data warehouse* is a subject-oriented, integrated, time-variant, non-updatable data set used in support of management decision-making processes and business intelligence. This definition is authored by Hackathorn.[3]

Data warehousing is the process that organizations employ to create and maintain data warehouses plus enabling extraction of information

---

[3] Hackathorn, R. (1993). *Enterprise Database Connectivity*. New York: Wiley.

useful for informed decision making. Data warehouses consolidate data located in disparate databases. By storing large quantities of data by categories, data warehouses enable efficient retrieval, interpretation, and storage of data. Data warehousing is one means of maintaining a central depository of all organizational data. This is the ideal software support needed for data mining.

Sometimes, in order to efficiently process certain applications, data mining is run on a subset of a data warehouse. In this case a ***data mart*** is employed. A data mart is a subset of the enterprise-wide database restricted to a single subject area. Data marts use a consistent data model and provide quality data.

2.5 EXPLORATORY DATA ANALYSIS AND VISUALIZATION

When performing a knowledge discovery task, visualization is a key ingredient at each step. Summary tables and statistics are useful for interpretation but are not comparable to displays of data points and their relationships. There is a standard collection of graph- and chart-drawing facilities common to all commercial discovery-oriented data mining tools. SAS and S statistical packages provide visualizations such as scatter-plots and multidimensional point cloud rotations.

Databases organized to support easy and efficient multidimensional analysis are referred to as *multidimensional databases*. When data mining a multidimensional database, *data cubes* provide for efficient retrieval of data. OLAP software for processing a cube feature page-by, pivot, sort, filter and drill-up, as well as, drill-down. These tools allow the end user to slice-and-dice a cube of data with mouse clicks.

Many OLAP vendors provide for three-dimensional visualization tools, which are applicable for data mining. Some of the new three-dimensional tools include dashboards and scorecards. Dashboards and scorecards provide visual displays of important information in a consolidated and organized manner on a single screen so that the information can be digested at a single glance and efficiently explored. Dashboards and scoreboards possess the following features:

- Use of visual components.
- Require minimal training.

- Data is derived from a variety of systems into a single summarized view.

- Enable drill-down or drill-through processing.

- Are based upon dynamic real-world view of timely data refreshes.

  Additionally, some vendors support tools for visual analysis. Data mining tools are accessible from the following sources:

- Mathematical and statistical packages.

- Web-based marketing packages.

- Analytics added to database tools.

- Standalone data mining tools.

  WEKA is an open source set of machine learning algorithms for data mining tasks. Neural network capabilities are included in WEKA. WEKA is downloadable from *cs.waikato.ac.nz/~ml/weka*.

2.6 DATA MINING ALGORITHMS

Statistical, machine learning, and neural network based algorithms are employed in data mining to capture classes, clusters, associations, and sequential patterns. These algorithms include components to:

- Post extracted, transformed, and transaction data needed to the data warehouse.

- Manage data in a multidimensional database.

- Provide multi-user data access, especially for business analysts and IT personnel.

- Perform data analysis.

- Provide graphical displays useful for interpretation.

  The data mining field is still evolving. Database personnel were among the first individuals who gave a serious thought to the problem of data mining, because they were the first to face the problem of extraction of hidden information in a database. Most of the tools and techniques used for data mining come from other related fields like pattern recognition, statistics, and complexity theory. Only recently have researchers from various fields been interacting to solve the mining issue.

Many of the traditional data mining techniques have failed because of the sheer size of the data. New techniques will have to be developed to store this huge data set. Any algorithm that is proposed for mining data will have to account for out of core data structures. Existing algorithms have not addressed the data set size issue. Some of the newly proposed algorithms, like parallel algorithms, are now beginning to look at data mining on large databases.

Most data mining algorithms assume the data to be noise free. As a result, the most time-consuming part of solving problems becomes data preprocessing. Data formatting and experimental/result management are frequently just as time consuming and frustrating.

2.7 MODELING FOR DATA MINING

Nonlinear analysis is provided by artificial neural networks, while algorithms like genetic algorithms offer optimization. Classification employs tree-structured algorithms. Rule induction involves the extraction of if-then rules from the data based upon statistical significance. Data visualization, which is extremely useful for data mining, provides visual interpretation of complex relationships in multidimensional data.

2.8 SUMMARY

Consider the following analogy. The patient, or the database, is examined for patterns that will extract information and rules applicable to the patient's cardiovascular system, or the enterprise-wide database and data warehouses. The skeletal system is the computer architecture and the tendons and muscle are the operating system and netware. Data mining algorithms, software, and methods are the data mining physician's black bag, or toolset. Through education and experience, the physician can generate useful information using his toolset. Like a medical doctor, beyond technical knowledge and know how, the data miner must know and understand his patient.

2.9 EXERCISES

1. Why is a standardized KDD necessary?

2. Compare data mining to the KDD process. Which term is broader?

3. Compare the two KDD processes discussed in this chapter.

4. Identify where mistakes can be made in data mining. For each potential mistake, discuss how to avoid them.

5. What factors have increased the popularity of data mining recently?

6. Identify at least five applications of data mining.

7. Identify some major characteristics of data mining.

8. Identify some of the main categories of data mining technologies.

9. Perform a topical search on the Six Sigma Method's approach to data mining.

10. Perform a topical search and prepare a report on text mining.

11. What is the relationship between OLAP and data mining?

12. Go to *http://www.teradastudentnetwork.com* and find seminars on data mining.

    a.   List and discuss applications of data mining.

    b.   Determine types of costs and payoffs that organizations can expect from data mining.

13. Discuss the major advantages of data warehousing to end users.

14. Distinguish between a data warehouse and a data mart.

15. How can data integration lead to higher levels of data quality?

16. How is a data warehouse different from a database?

17. Why has data mining suddenly gained the attention of the business world?

18. Describe the algorithm and provide example applications for the data mining functions.

19. Explain how pivot tables can be employed in data mining.

HIERARCHICAL CLUSTERING

In This Chapter

3.1 INTRODUCTION

Hierarchical clustering can be broken down into two major categories—*agglomerative methods* and *divisive methods*. A procedure for forming agglomerative hierarchical groups of mutually exclusive subsets was developed by Ward.[1] The grouping technique is "based on the premise that the greatest amount of information, as indicated by the objective function, is available when a set of n members is ungrouped."[1] The first step is to select and combine the two subsets out of the n possible subsets which would produce the least impairment of the optimum value of the objective function, while

---

[1] Ward, J. H., Jr. (1963). Hierarchical grouping to optimize an objective function. *American Statistical Association Journal*, 58, 236-244.

reducing the number of subsets to $n - 1$. These $n - 1$ subsets are then examined to identify which two subsets should be merged in order to obtain the optimum value of the objective function for $n - 2$ subsets. This procedure is repeated until the original n members are in a single set or group. Because the number of subsets is reduced by one at each step $(n - 1,..., 1)$, the process is referred to as *hierarchical grouping*. The grouping that occurs at each step usually results in some quantifiable loss, which Ward[2] terms a *value-reflecting* number. The functional relation used to obtain a value-reflecting number is called the *objective function*. In general, an objective function may be any functional relationship that an investigator selects to reflect the relative desirability of groupings. In Ward's[1] example, the objective function represents loss of information and is reflected by the error sum of squares (now known as Ward's Method). Other examples of objective functions described by Ward[1] which have been used by the Personnel Research Laboratory, United States Air Force, include:

1. the grand sum of the squared deviations about the means of all measured characteristics in the clustering of persons to maximize their similarity with respect to measured characteristics,

2. the expected cross-training time in the clustering of jobs to minimize cross-training time when personnel are reassigned according to established policies,

3. the amount of job time incorrectly described when clustering job descriptions to minimize errors in describing a large number of jobs with a small number of descriptions, and

4. the loss of predictive efficiency when clustering regression equations to minimize the loss of predicted efficiency resulting from reductions in the number of regression equations (JAN).

An important concept to remember in defining the hierarchical grouping mentioned previously is that once two entities are merged, they are joined permanently and become a building block for later merges. This is one of the key differences between the hierarchical techniques and the optimal techniques to be discussed in Chapter 4. Another difference between

---

[2] Ward, J. H, Jr. (1961, March). Hierarchical grouping to maximize payoff (*Technical Report WADD-TN-61-29*). Lackland Air Force Base, TX: Personnel Laboratory, Wright Air Development Division.

the two methods is how they arrive at the number of groups. If researchers are not interested in the entire hierarchical tree structure, they must decide at what point the "best" clustering has been reached. There are many different techniques available to help the user make a selection. See Anderberg[3] and Everitt[4] for more information on deciding on the number of clusters. Some of the more common hierarchical techniques are the nearest neighbor (single linkage) method, the furthest neighbor (complete linkage) method, centroid method, and Ward's method (mentioned previously). The basic technique for these methods is similar. However, there are two factors that differentiate between the methods. The first factor is whether the technique can be used with similarity measures, distance measures, or both. The second factor is how these measures are defined, or, in Ward's terminology, the objective function. Detailed information on these methods and the methods to be discussed later can be found in a variety of cluster analysis or multivariate texts represented by Anderberg[3]; Everitt[4]; Hartigan[5]; Jain & Dubes[6]; and Seber.[7]

In divisive clustering, the initial starting point is with all entities in one group. The first step is to split the group into two subsets. Subsequent steps consist of subdividing the subsets until each entity is a separate group. There are two classes of methods for deciding how to split the groups. With the homothetic methods, the data points are based on binary variables and the object is to split the data set on one of these variables so as to minimize the value of some appropriate measure of similarity between the two groups. The splitting process, based on other variables, continues until some criterion is reached as discussed in Anderberg.[3] Examples of homothetic techniques include association analysis and the automatic interaction detector method. The polythetic methods are based on the values taken by all of the variables. According to Everitt,[4] a technique developed by MacNaughton-Smith, Williams, Dale, and Mockett is the most feasible. The divisive techniques have not gained the same popularity as the agglomerative techniques and their use is minimal.

---

[3] Anderberg, M. R. (1973). *Cluster analysis for applications*. New York: Academic Press.

[4] Everitt, B. S. (1980). *Cluster analysis (2nd ed.)*. New York: Halsted Press.

[5] Hartigan, J. A. (1975). *Clustering Algorithms*. New York: John Wiley and Sons.

[6] Jain, A. K., & Dubes, B. C. (1973). *Algorithms for clustering data*. Englewood Cliffs, NJ: Prentice Hall.

[7] Seber, G. A. F. (1984). *Multivariate observations*. New York: John Wiley and Sons.

3.2 SINGLE-LINK VERSUS COMPLETE-LINK CLUSTERING

Recall the cluster analysis from Chapter 1, a complete-link clustering. The following results were obtained:

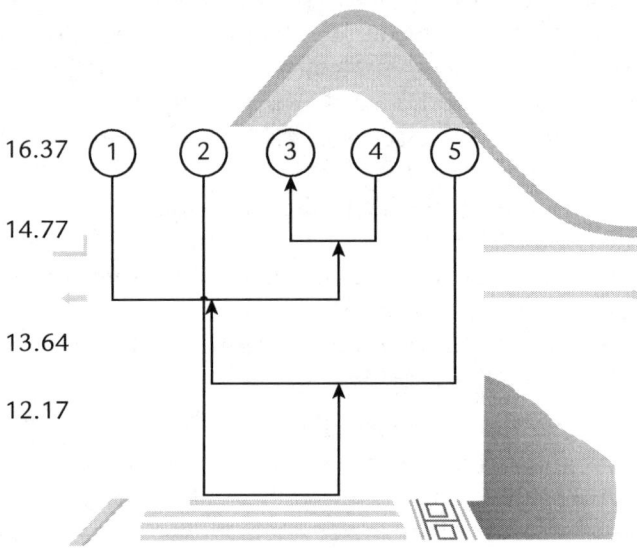

FIGURE 3.1 Complete-Link Clustering.

The dendrogram in Figure 3.1 was generated by the following tables as illustrated in Table 3.1. A new matrix is formed by deleting the column and the row of the cell containing the maximum value. The deleted row and deleted column represent the identifier for the new pseudo data point. The distance from the pseudo data point to a specified data point is the maximum of the distance between the two data points, designated by the original row and original column for the pseudo data point, to the specified data point.

The single-link clustering is obtained by the same process as the complete-link computations except minimum values are utilized rather than maximum as illustrated in Figure 3.2.

Figure 3.2 is based upon the ratio proximity computations in Table 3.2.

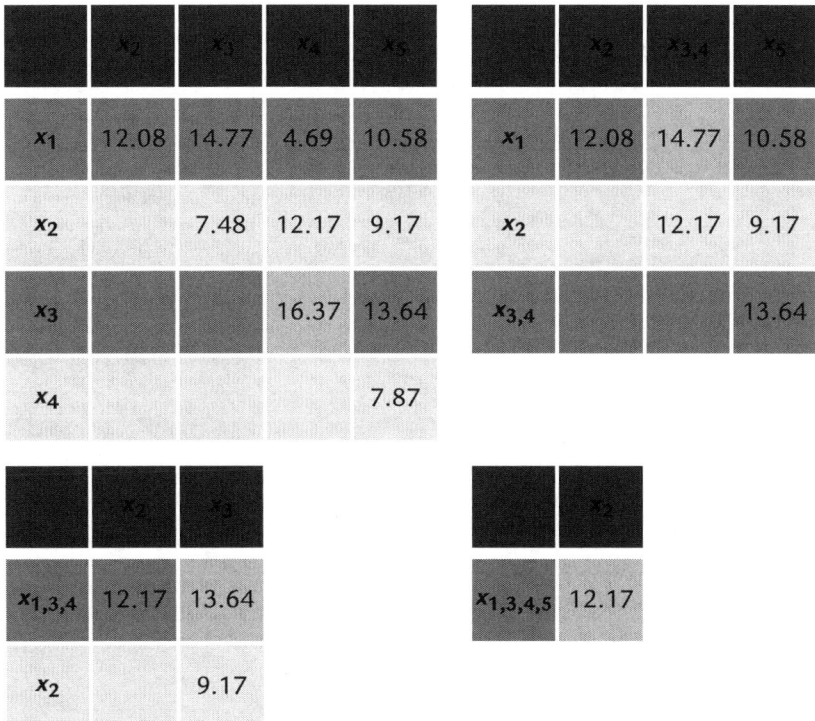

	x_2	x_3	x_4	x_5
x_1	12.08	14.77	4.69	10.58
x_2		7.48	12.17	9.17
x_3			16.37	13.64
x_4				7.87

	x_2	$x_{3,4}$	x_5
x_1	12.08	14.77	10.58
x_2		12.17	9.17
$x_{3,4}$			13.64

	x_2	x_3
$x_{1,3,4}$	12.17	13.64
x_2		9.17

	x_2
$x_{1,3,4,5}$	12.17

TABLE 3.1 Complete-Link Clustering Matrix Computations.

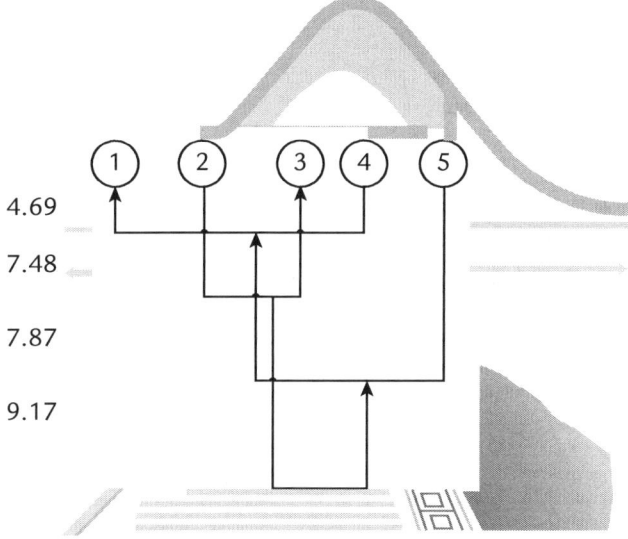

FIGURE 3.2 Single-Link Clustering.

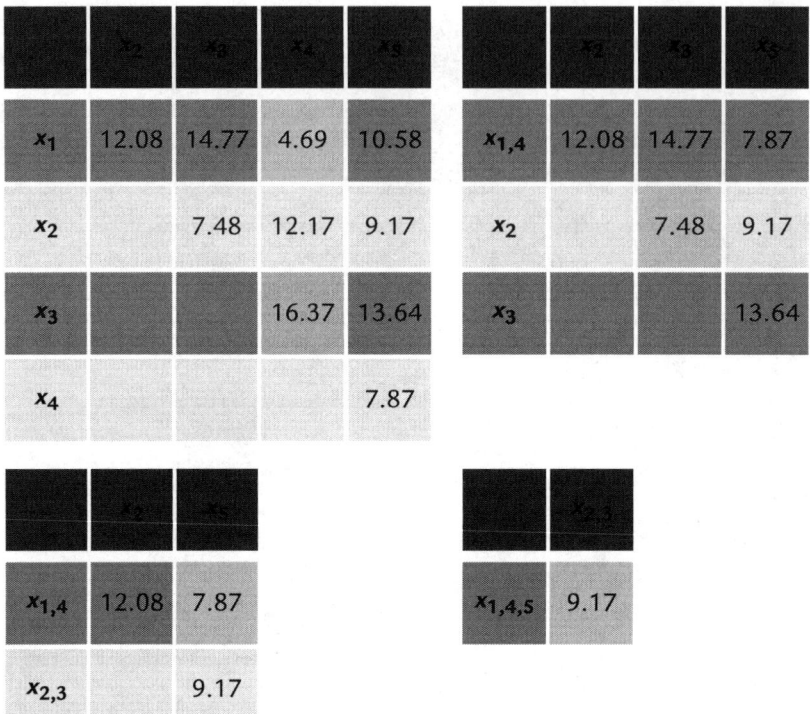

	x_2	x_3	x_4	x_5
x_1	12.08	14.77	4.69	10.58
x_2		7.48	12.17	9.17
x_3			16.37	13.64
x_4				7.87

	x_2	x_3	x_5
$x_{1,4}$	12.08	14.77	7.87
x_2		7.48	9.17
x_3			13.64

	x_2	x_5
$x_{1,4}$	12.08	7.87
$x_{2,3}$		9.17

	$x_{2,3}$
$x_{1,4,5}$	9.17

TABLE 3.2 Single-Link Clustering Matrix Computations.

Note that these methods generate distinct clustering sets:

$$\{(x_1), (x_2), (x_3), (x_4), (x_5)\} \text{ at the start}$$
$$\{(x_1), (x_2), (x_3, x_4), (x_5)\} \text{ at } L(1) = 16.37$$
$$\{(x_2), (x_1, x_3, x_4), (x_5)\} \text{ at } L(2) = 14.77$$
$$\{(x_1, x_3, x_4, x_5,), (x_2)\} \text{ at } L(3) = 13.14$$
$$\{(x_1, x_2, x_3, x_4, x_5)\} \text{ at } L(4) = 12.17$$

FIGURE 3.3 Complete-Link Clustering Iterations.

$$\{(x_1), (x_2), (x_3), (x_4), (x_5)\} \text{ at the start}$$
$$\{(x_1, x_4), (x_2), (x_3), (x_5)\} \text{ at } L(1) = 4.69$$
$$\{(x_1, x_4), (x_2, x_3), (x_5)\} \text{ at } L(2) = 7.48$$
$$\{(x_1, x_4, x_5), (x_2, x_3)\} \text{ at } L(3) = 7.87$$
$$\{(x_1, x_2, x_3, x_4, x_5)\} \text{ at } L(4) = 9.67$$

FIGURE 3.4 Single-Link Clustering Iterations.

When should each of these clustering iterations be stopped? An intuitive answer would be based upon the values for L(m). A first step would be to plot the number of groups versus L(m) and look for points where the curve flattens or at the position where the curve has a "knee" or "elbow."

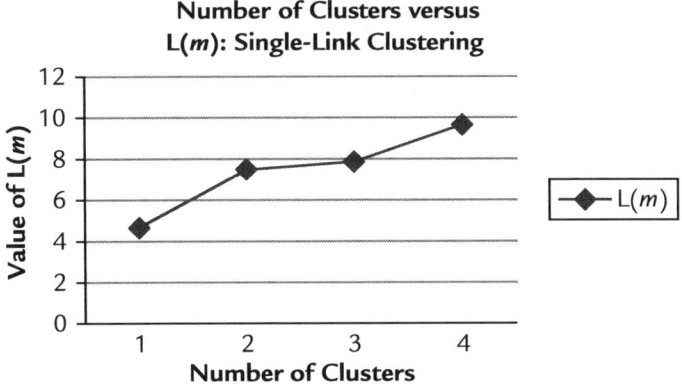

FIGURE 3.5 One Cluster Single-Link Solution.

For example, instead of a one cluster solution, the researcher could look for a bend in the curve. This suggests either a two cluster solution, $\{(x_1, x_4), (x_2, x_3), (x_5)\}$, or a three cluster solution, $\{(x_1, x_4, x_5), (x_2, x_3)\}$, for the single-link analysis.

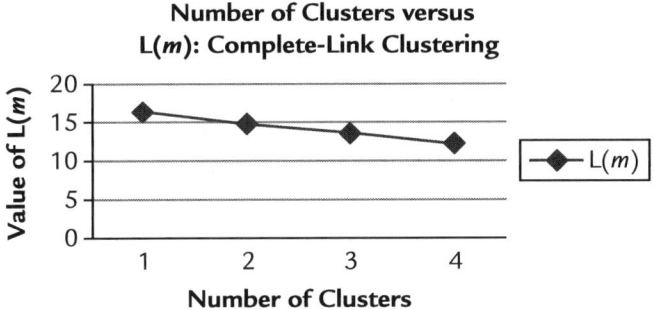

FIGURE 3.6 One Cluster Solution for Complete-Link Clustering.

For the complete-link clustering analysis there is a one cluster solution, $\{(x_1, x_2, x_3, x_4, x_5)\}$, when applying either the flatten curve or curve-bending method to determine the number of clusters. Clearly the question, "How many clusters?", is difficult to answer. The researcher needs a way to consider merging costs.

Often information is inaccessible for determining the right number of clusters, let alone understanding what is meant by the right number of clusters. If every object is a cluster, then the total information about the data is present. But then why perform a cluster analysis? Another obvious clustering choice is to choose the one cluster solution, then misrepresentation is possible, unless there actually is only one cluster. The emphasis in cluster analysis is to generate models which generalize to new data. If the data really does fall into k clusters, then more data from the same source should fall into the same clusters. There are multiple ways for measuring this, based upon redoing the analysis by measuring cluster efficiency, which is available when using Ward's method or a K-Means approach.

1. How much do cluster centers or boundaries change if we rerun clustering on new data?

2. How much do cluster assignments change, during an optimization of the existing clustering?

3. How big are the sum-of-squares when assigning new data to the old clusters?

4. What is the metric distance between old and new clusterings?

Either new data is required or reusing some of the current data can answer these questions.

A distinct but related question on how many clusters are representative of the data is: "What is the confidence in the statement: 'these two data points belong to the same cluster'?" First a cross-validation analysis should be performed, followed by an examination to determine how often the points in question are posted in the same cluster. In general:

1. Clusters should continue to describe new observations of the same features.

2. Good clusters should enable the researcher to generate new features.

3. Good clusters should be a part of a valid system which enables making predictions about new conditions.

4. A valid cluster membership system should provide explanations for the experimental results.

3.3 AGGLOMERATIVE VERSUS DIVISIVE CLUSTERING

Divisive hierarchical clustering starts with the complete data set considered as a single-cluster, and splits it into two child-clusters. The process is iterated recursively until there is nothing but singletons left. The same dendrogram and selection procedures that are used for agglomerative hierarchical clustering may also be used for the divisive approach.

Agglomerative and divisive hierarchical clustering are not mirror images of each other because:

1. The agglomerative method requires that a table of pair-wise distances of clusters be updated after each merge. Choosing the next merge consists of finding the smallest distance in this table. Each step in the divisive method requires that a cluster be scrutinized for splitting. Maintaining a strict splitting criterion, requires that all possible 2-splits of the cluster be considered. Then the best choice is retained, resulting in a method that becomes unmanageably long even for modest size clusters. A reasonable approach is to forget about finding the best split, using shortcuts, and be satisfied with identifying just a "good" split.

2. The divisive method usually provides clusters that are easier to interpret than clusterings produced by the agglomerative method. The reason is that being close according to Ward's criterion, discussed in the next section, can be interpreted in terms of cluster geometry. Whereas partitioning a cluster into two clusters with a large Ward distance generates two child-clusters that are compact and well separated. The divisive method is closer to the way the human eye does clustering than the agglomerative method.

In practice, the difficulties associated with the divisive method make the agglomerative method more widely used.

3.4 WARD'S METHOD

Ward's method is associated with analysis of variance, instead of using distance metrics or measures of association, using an error measurement at each iterative step. This method starts out at the leaves and work its way to the trunk of the related dendrogram. Ward's method starts out with n clusters of size 1 and continues until all the observations are included into one cluster. This method is most appropriate for quantitative variables, and not binary variables.

Ward's method is based on the notion that clusters of multivariate observations should be approximately elliptical in shape. Therefore, it would follow that points would fall into an elliptical shape when plotted in a p-dimensional scatter plot.

The notation employed follows: Let X_{ijk} denote the value for variable j in observation i belonging to cluster k. Furthermore, for this particular method, we have to define the following error sum of squares (ESS):

$$\text{ESS} = \sum_i \sum_j \sum_k |x_{ijk} - \overline{x}_{i.k}|^2$$

where for a given classification observation i belongs to cluster k, x_{ij} is the value of variable j for observation i, and $\overline{x}_{i.k}$ is the mean of variable j in cluster k. The summation process includes all of the units within each cluster. A comparison of the individual observations for each variable against the cluster means for that variable is calculated. A small value for ESS indicates that data is close to the cluster means, implying clusters of similar units.

The total sums of squares (TSS) is defined as:

$$\text{TSS} = \sum_i \sum_j \sum_k |x_{ijk} - \overline{x}_{..k}|^2$$

again where for a given classification observation i belongs to cluster k, x_{ij} is the value of variable j for observation i, and $\overline{x}_{..k}$ is the grand mean in cluster k. The summation process includes all of the units within each cluster. A comparison of the individual observations for each variable against the grand mean for that variable is computed.

R-squared, $r^2 = \dfrac{TSS - ESS}{TSS}$, is interpreted as the proportion of variation explained by a particular clustering of the observations.

Therefore, the summations are to find the squared error of observation i in cluster k for all the variables, for all observations i within cluster k, and to find the total error over all clusters k.

Ward's method starts with all sample units in n clusters of size one. In the first step of the algorithm, $n - 1$ clusters are formed, one of size two and the remaining of size one. The error sum of squares and r^2 values are then computed. The pair of sample units that yield the smallest error sum of squares, or equivalently, the largest r^2 value will form the first cluster. Then, in the second step of the algorithm, $n - 2$ clusters are formed from that $n - 1$ clusters as defined in step 2. These may include two clusters of size 2, or a single cluster of size 3 including the two items clustered in step 1. Again, the value of r^2 is maximized. Thus, at each step of the algorithms, clusters or observations are combined in such a way as to minimize the results of error from the squares or alternatively to maximize the r^2 value. The algorithm stops when all sample units are combined into a single large cluster of size n.

Consider the following data file:

ID	Gender	Age	Salary
1	F	27	19,000
2	M	51	64,000
3	M	52	100,000
4	F	33	55,000
5	M	45	45,000

TABLE 3.3 Original Sample Employee Data Table.

First, each variable must be quantitative, therefore, change the data in Table 3.3 into the table found in Table 3.4.

ID	Gender	Age	Salary
1	1	0.00	0.00
2	0	0.96	0.56
3	0	1.00	1.00
4	1	0.24	0.44
5	0	0.72	0.32

TABLE 3.4 Fully Quantitative Version of Table 3.3.

A sample set of computations for the square error of all IDs for all pairs of object includes:

$$\text{dist}((ID2), (ID1)) = \text{SQRT}(1 + (.096)^2 + (0.56)^2) = 2.24$$

$$\text{dist}((ID2), (ID3)) = \text{SQRT}(0 + (0.04)^2 + (0.44)^2) = 0.44$$

$$\text{dist}((ID2), (ID4)) = \text{SQRT}(1 + (0.72)^2 + (0.12)^2) = 1.24$$

$$\text{dist}((ID2), (ID5)) = \text{SQRT}(0 + (0.24)^2 + (0.24)^2) = 1.15$$

$$\text{dist}((ID1), (ID3)) = \text{SQRT}(1 + (-1.00)^2 + (1.00)^2) = 3.00$$

$$\text{dist}((ID1), (ID4)) = \text{SQRT}(0 + (-0.24)^2 + (-0.44)^2) = 0.25$$

$$\text{dist}((ID1), (ID5)) = \text{SQRT}(1 + (-0.72)^2 + (-0.32)^2) = 1.62$$

$$\text{dist}((ID3), (ID4)) = \text{SQRT}(1 + (0.76)^2 + (0.56)^2) = 1.89$$

$$\text{dist}((ID3), (ID5)) = \text{SQRT}(0 + (0.28)^2 + (0.68)^2) = 0.54$$

$$\text{dist}((ID4), (ID5)) = \text{SQRT}(1 + (-0.48)^2 + (0.12)^2) = 1.24$$

which, on the first iteration of Ward's method, generates the new clustering, {(ID1, ID4), (ID2), (ID3), (ID5)} at L(1) = 0.25 from the original cluster {(ID1), (ID2), (ID3), (ID4), (ID5)}.

For the second iteration, perform the following computations and obtain the resultant clustering, using the revised data table.

Cluster	Gender	Age	Salary
((ID1), (ID4))	0	0.12	0.22
(ID2)	0	0.96	0.56
(ID3)	0	1.00	1.00
(ID4)	1	0.24	0.44

TABLE 3.5 Second Iteration Table Using Ward's Method.

$$\text{dist}(((ID1), (ID4)), (ID2)) = \text{SQRT}(0 + (-0.84)^2 + (-0.33)^2) = 0.82$$

$$\text{dist}(((ID1), (ID4)), (ID3)) = \text{SQRT}(0 + (-0.88)^2 + (-0.78)^2) = 1.38$$

$$\text{dist}(((ID1), (ID4)), (ID5)) = \text{SQRT}(1 + (-0.12)^2 + (-0.22)^2) = 1.86$$

$$\text{dist}((ID2), (ID3)) = \text{SQRT}(0 + (-0.04)^2 + (-0.44)^2) = 0.20$$

$$\text{dist}(((ID2), (ID5)) = \text{SQRT}(1 + (-0.72)^2 + (-0.12)^2) = 1.53$$

$$\text{dist}((ID3), (ID5)) = \text{SQRT}(1 + (0.76)^2 + (0.56)^2) = 1.24$$

This generates the clustering, {(ID1, ID4), (ID2, ID3), (ID5)} at L(2) = 0.20.

On the third iteration, based upon the following table, the results are:

Cluster	Gender	Age	Salary
((ID1), (ID4))	0	0.12	0.22
((ID2), (ID3))	0	0.98	0.78
(ID5)	1	0.24	0.44

TABLE 3.6 Third Iteration Table Using Ward's Method.

dist(((ID1), (ID4)), ((ID2), (ID3))) = SQRT(0 + (–0.86)^2 + (–0.56)^2) = 1.05

dist(((ID1), (ID4)), (ID5) = SQRT(1 + (–0.12)^2 + (–0.22)^2) = 1.06

dist(((ID2), (ID3)), (ID5)) = SQRT(1 + (–0.72)^2 + (0.34)^2) = 1.63

which generates the clusterings {(ID1, ID4), (ID2, ID3), (ID5))} with L(3) = 1.05. Note, in this example the clustering could have been {(((ID1), (ID4)), (ID5)), ((ID2), (ID3))} instead.

On the fourth generation we obtain:

Cluster	Gender	Age	Salary
(((ID1), (ID4)), ((ID2), (ID3)))	0	0.55	0.50
(ID5)	1	0.24	0.44

TABLE 3.7 Fourth Iteration Table Using Ward's Method.

dist((((ID1), (ID4)), ((ID2), (ID3))), (ID5)) = SQRT(1 + (0.31)^2 + (0.06)^2) = 1.10 which generates the clustering {ID1, ID2, ID3, ID4, ID5} where L(4) = 1.10.

3.5 GRAPHICAL ALGORITHMS FOR SINGLE-LINK VERSUS COMPLETE-LINK CLUSTERING

Consider the proximity data matrix in Table 3.8:

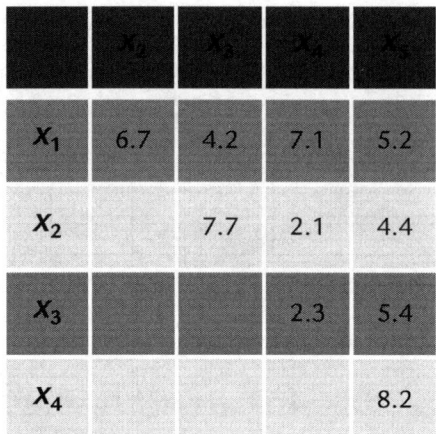

	X_2	X_3	X_4	X_5
X_1	6.7	4.2	7.1	5.2
X_2		7.7	2.1	4.4
X_3			2.3	5.4
X_4				8.2

TABLE 3.8 Proximity Table.

A picture of the data is needed that can be easily interpreted. Dendrograms offer a partial solution because they list the clustering and a horizontal cut of a dendrogram defines and identifies clusters, but not the actual dissimilarity values for which clusters are formed. Because any hierarchical clustering scheme is simply a way of transforming a proximity matrix into a dendrogram, then a solution to the latter problem is to construct a sequence of proximity graphs.

Step 1: Represent the proximity as a graph where nodes represent the objects, x_i for $I = 1$ to 5. The edges are the minimum distance between the nodes and, simultaneously when recorded, are the least similar pair of edges in the proximity graph.

Step 2: Repeat Step 1 until the edges are exhausted.

For the given proximity matrix the following sequence of graphs is generated:

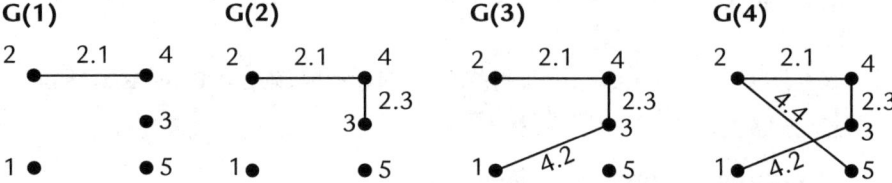

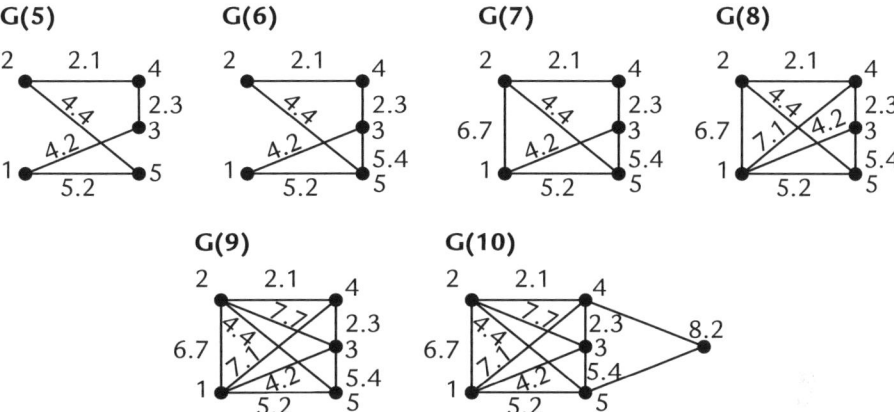

FIGURE 3.7 Threshold Graphs.

These graphs generate threshold dendrograms. Figure 3.6 illustrates the single-link clustering with the threshold dendrogram. Start with the proximity matrix and on each iteration apply the minimum distance to all clusters formed.

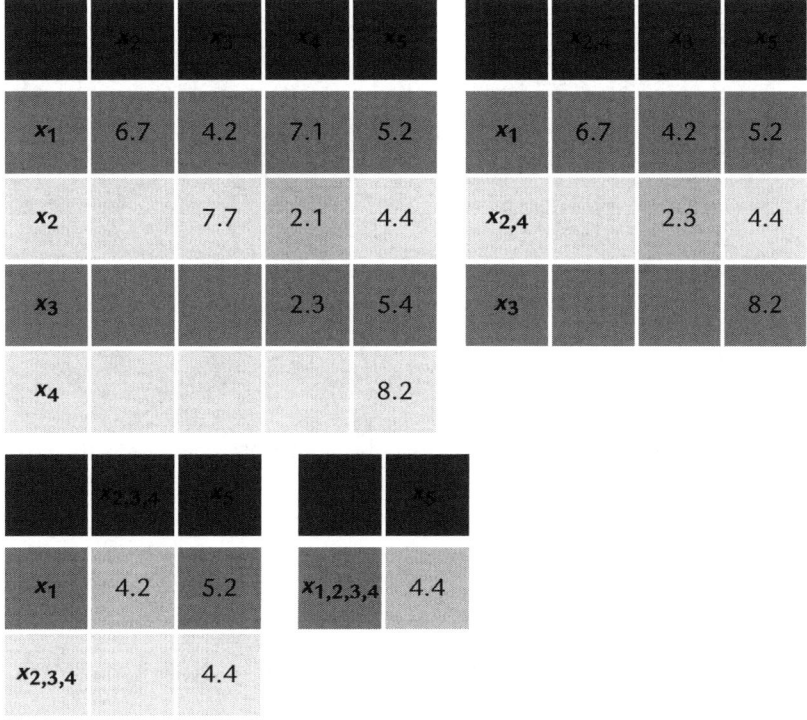

TABLE 3.9 Single-Link Clustering Process.

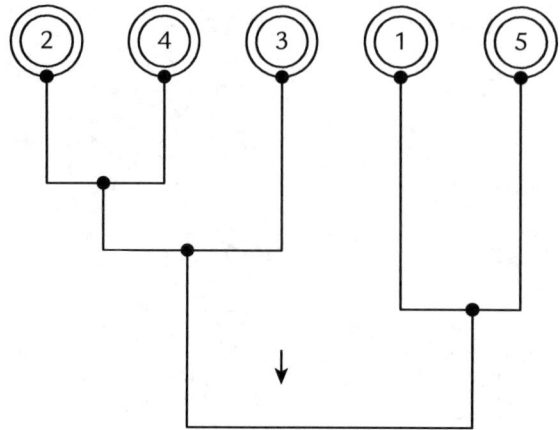

FIGURE 3.8 Threshold Dendrogram: Single-Link Clustering.

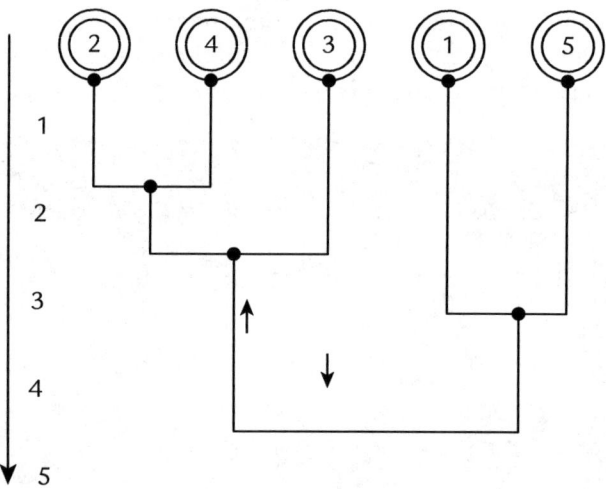

FIGURE 3.9 Proximity Dendrogram for Single-Link Clustering.

A **proximity dendrogram** posts both the proximities and the proximities cluster formation. The results indicate a two cluster solution: $\{(x_2, x_3, x_4), (x_1, x_5)\}$.

3.6 SUMMARY

- Hierarchical cluster analysis is a very popular clustering paradigm that includes several different methods.

■ Given a data set, a **hierarchy** is a set of subsets of data (clusters) such that:

   1. One cluster contains the complete set of observations.

   2. Another cluster contains all the singleton observations.

   3. For every pair of clusters in the hierarchy, the pair of clusters either has an empty intersection or else one cluster is included in the other cluster (no overlapping).

   4. Any cluster (except the singletons) is partitioned into exactly two clusters of the hierarchy.

■ The structural form for a clustering solution graphically is represented by a **Dendrogram** (or tree).

■ Hierarchical agglomerative clustering methods using the agglomerative approach appear to have the advantage of finding any number of compact, spherically shaped clusters at a certain level which cover the sample density. Divisive hierarchical and agglomerative hierarchical clustering methods employ "divide and conquer" algorithms.

■ Ward's method performs in a manner similar to the decision tree approach, namely it combines the two clusters at each stage which minimize the squared error function or Euclidean sum of squares. Ward's method uses a greedy algorithmic analysis.

■ Requirements for any clustering method should:

   1. Have scalability

   2. Deal with different types of attributes

   3. Discover clusters with arbitrary shape

   4. Have minimal requirements for domain knowledge to determine input parameters

   5. Be able to deal with noise and outliers

   6. Be insensitive to order of input records

   7. Have the curse of dimensionality

   8. Have interpretability and usability

■ Common agglomerative hierarchical clustering methods are:

- **Single-link or nearest neighbor method**: where the dissimilarity between two clusters is equal to the minimum of all distances between the cases in the participating clusters.

- **Complete-link or farthest neighbor method**: the dissimilarity between two clusters is defined to be the maximum for all possible distances between the cases in the two participating clusters.

- **Ward's method**: where the dissimilarity between two clusters is defined to be the loss of information from joining the two clusters. Loss of information is found by measuring the increase in the error sum of squares, or the sum of squared deviations of each pattern from the centroid for the cluster.

- **Group average (mean) method**: uses the average value of pair wise links within a cluster to determine intercluster similarity, because all objects contribute to intercluster similarity.

- **Centroid method**: where the dissimilarity between two clusters is defined by the respective centroids. The squared Euclidean distance instead of the Euclidean should be used as the dissimilarity when applying the centroid method to a data set.

- **Median method**: where the dissimilarity between clusters is defined by the distance between the medians for the two clusters. Again the Euclidean squared distance instead of the Euclidean distance should be used for this method.

When performing a cluster analysis, the investigator should apply several of these methods to the data and seek the majority resultant clustering across the methods as the final clustering chosen. For example, consider the following data {(12, 3), (16, 19), (23, 13), (20, 23), (8, 20), (16, 9), (1, 23), (25, 20), (18, 6), (5, 28)} with associated labels {1, 2, 3, 4, 5, 6, 7, 8, 9, 10}, which are points in two dimensional Euclidean space, these methods generate the results below:

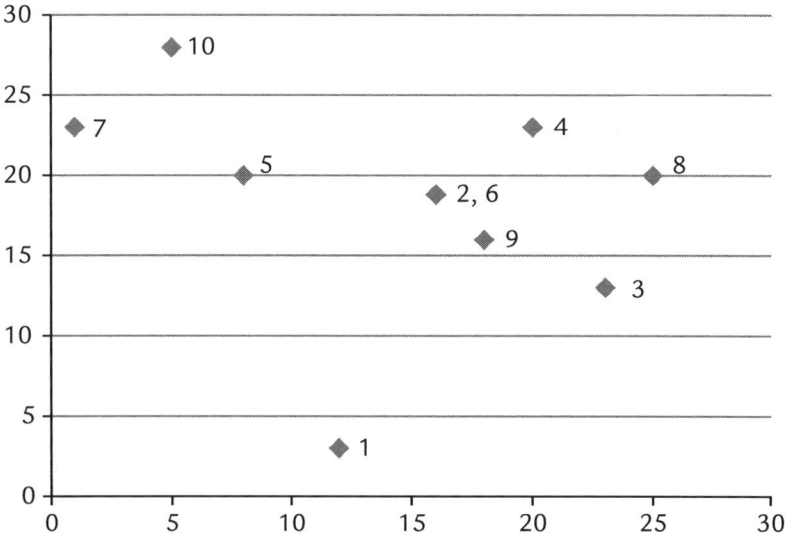

FIGURE 3.10 Sample Study Data Set.

Clusterings are obtained using several of the hierarchical clustering methods and followed by obtaining the majority clustering final solution.

NOTE: All figures and dendrograms in the rest of this Chapter were generated by Wessa[8], using the R framework system

[8] Wessa, P., (2008). Hierarchical Clustering (v1.o.2) in Free Statistics Software (v1.1.23-r7) office for Research and Development and Education, URL http://www.wessa.net/rwasp-hierarchicalclustering.wasp/.

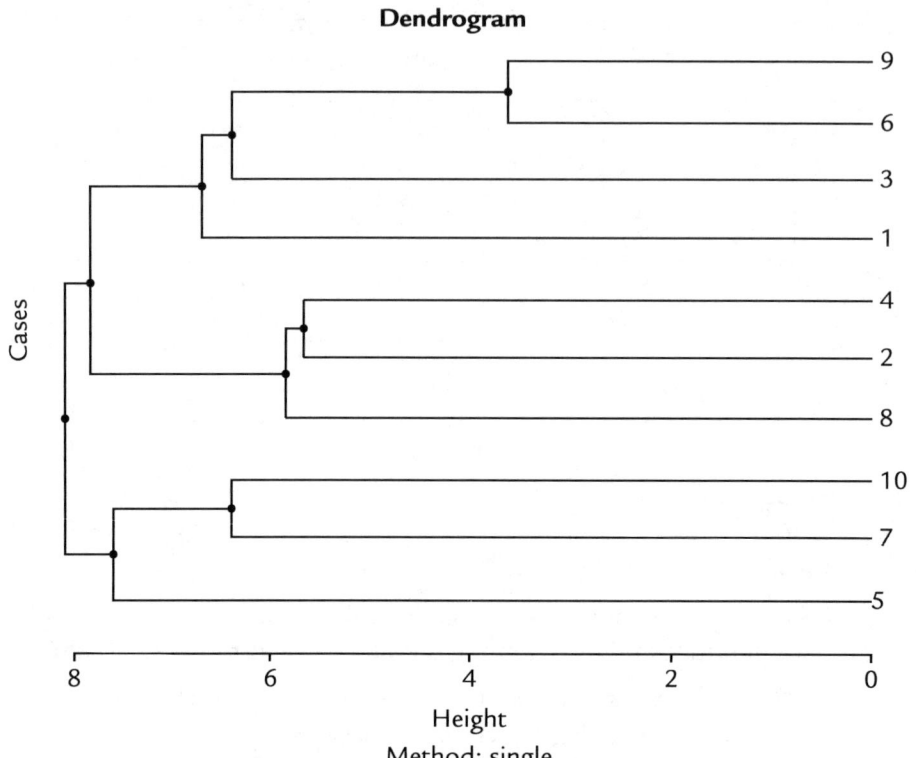

Clustering: {{3, 6, 9, 1}, {4, 2 8}, {7, 5, 10}}

FIGURE 3.11 Single-Link Clustering for the Sample Study.

Summary of Dendrogram	
Label	Height
1	3.60555127546399
2	5.65685424949238
3	5.8309518948453
4	6.40312423743285
5	6.40312423743285
6	6.70820393249937
7	7.61577310586391
8	7.81024967590665
9	8.06225774829855

Dendrogram

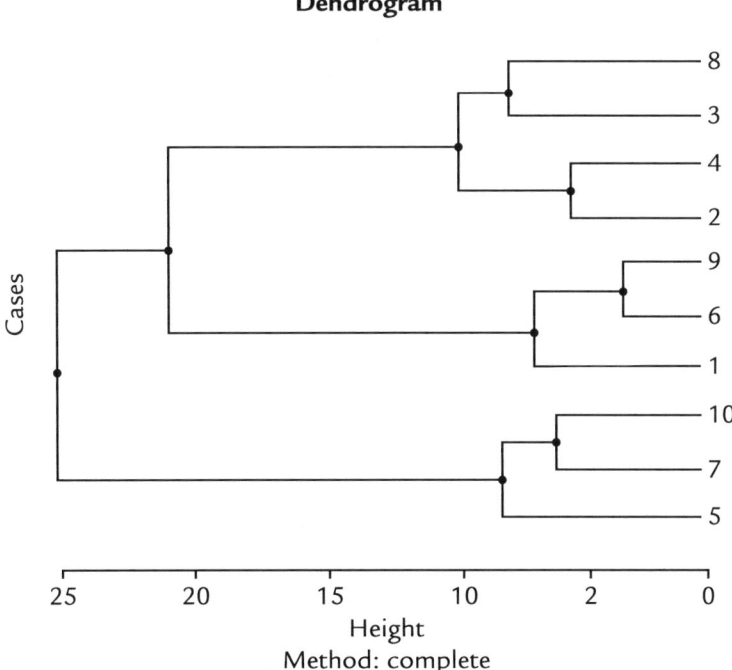

Method: complete

Summary of Dendrogram	
Label	Height
1	3.60555127546399
2	5.65685424949238
3	6.40312423743285
4	7.21110255092798
5	8.06225774829855
6	8.54400374531753
7	10.0498756211209
8	21.540659228538
9	25.9615099714943

Clustering: {{2, 4, 3, 8}, {1, 6, 9}, {5, 7, 10}}

FIGURE 3.12 Complete-Link Clustering for the Sample Study.

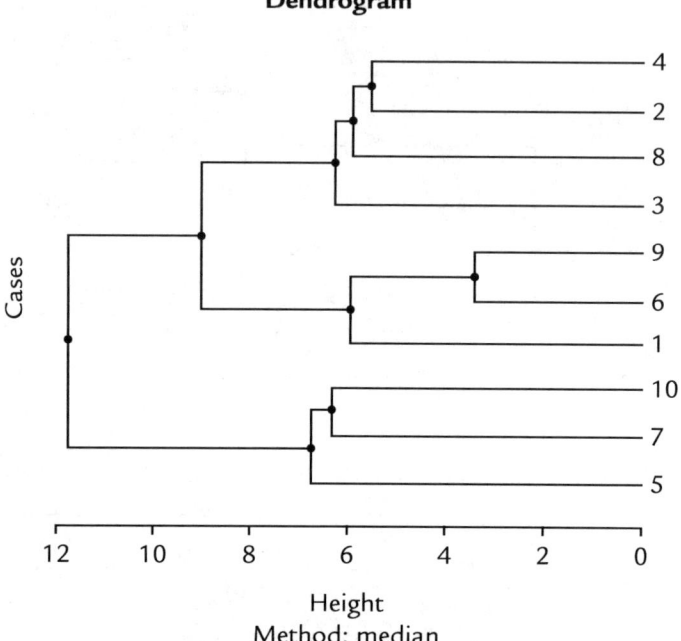

Summary of Dendrogram	
Label	Height
1	3.60555127546399
2	5.65685424949238
3	6.02895495411826
4	6.05826542284768
5	6.28181467869005
6	6.40312423743285
7	6.47910736623251
8	9.10508783423627
9	11.7692606295293

Clustering: {{2, 3, 4, 8}, {1, 6, 9}, {5, 7, 10}}

FIGURE 3.13 Median Clustering for the Sample Study.

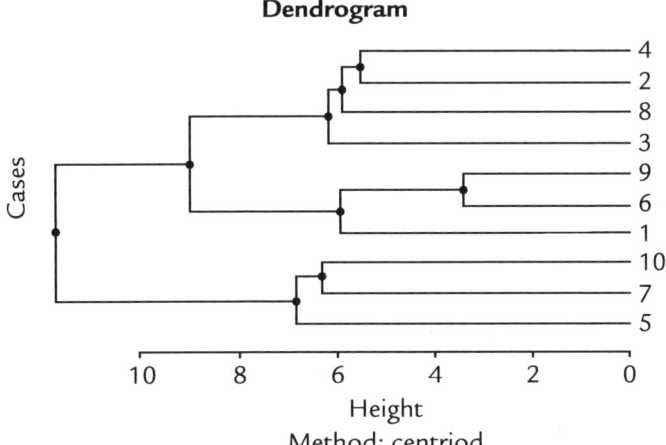

FIGURE 3.14 Centroid Clustering for the Sample Study.

Final Clustering: {{1, 6, 9}, {2, 3, 4, 8}, {5, 7, 10}}

NOTE: 2 and 6 are identical points.

The following discussion is a sample study of sixty-three software companies involved in software development projects. Identifying how many clusters of these companies exist will be pertinent to eventually making an effective assessment comparison of the companies common variates for estimating software development costs. Once the clusters are identified then the cluster centroids can be used to represent the clusters. Then a multiple regression analysis and predictive study should be performed on each cluster individually.

t01	t02	t03	t04	t05	t06	t07	t08	t09	t10	t11	t12	t13	t14	t15
5	3	3	2	3	4	4	4	3	3	2	4	3	5	3
4	3	5	3	3	3	4	5	4	5	4	4	4	4	5
2	3	3	3	3	3	2	2	4	3	4	4	4	4	4
3	3	2	3	3	4	2	3	4	5	4	3	2	3	3
2	2	4	2	2	1	3	5	4	4	5	4	3	2	3
3	3	3	4	3	3	4	3	4	4	3	4	5	4	4
2	3	3	3	3	3	2	2	4	4	4	4	4	5	4
4	3	5	4	3	2	3	5	5	5	3	4	4	2	3
2	3	3	2	2	2	4	5	4	3	3	3	3	2	3
4	3	3	2	1	2	4	5	3	2	2	2	3	4	2
2	3	2	3	3	3	2	5	3	4	2	3	2	3	3
5	3	4	2	3	1	3	3	3	2	2	2	1	1	2
2	2	2	4	3	3	1	4	4	3	4	4	1	5	1
2	3	3	4	2	2	2	4	4	3	5	3	3	4	2
2	3	2	4	3	3	3	5	4	3	3	4	2	4	3
3	4	3	3	3	3	3	3	5	5	2	4	3	3	3
2	3	2	4	3	4	4	4	4	3	2	4	3	3	3
4	1	3	3	3	4	4	5	4	4	4	4	3	3	4
4	3	4	3	4	4	5	4	5	4	5	5	3	1	4
3	2	3	3	3	2	4	5	4	4	4	4	4	2	3
4	3	4	4	4	2	3	4	5	5	3	4	4	2	3
4	4	2	3	3	3	4	3	4	4	5	3	2	3	3
4	3	2	4	3	3	4	4	5	4	3	4	2	4	3
4	3	2	3	2	3	2	4	4	4	2	2	3	3	3
2	2	2	4	3	3	4	4	5	5	5	4	2	2	4
4	3	4	3	3	3	3	3	4	3	3	3	2	3	3
4	3	3	4	3	3	4	5	4	4	4	4	4	2	4
3	3	3	4	3	4	2	4	4	3	2	4	4	2	4
1	4	2	3	3	4	2	3	2	3	2	4	2	4	3
3	4	2	3	3	4	3	5	3	3	3	4	4	5	3
2	4	3	2	2	2	3	4	5	4	4	4	3	3	4
1	3	3	3	3	2	4	5	3	2	3	3	3	4	3
3	4	2	3	3	3	3	4	3	4	2	4	3	3	3

3	3	3	4	3	3	2	3	4	2	3	4	2	4	3
4	2	4	3	5	3	4	5	5	4	3	5	4	4	5
4	2	4	3	5	3	5	3	5	5	4	5	3	4	5
4	2	4	3	5	3	5	3	5	5	4	5	3	4	5
3	3	4	2	3	2	3	4	3	3	3	4	4	3	3
4	3	4	4	4	3	3	4	4	4	5	5	4	3	4
2	3	2	4	3	3	3	4	3	3	3	3	2	4	3
4	2	2	3	4	4	4	2	4	2	4	3	4	3	4
3	3	3	2	2	3	3	3	4	3	3	4	4	4	3
3	4	2	4	3	3	3	3	5	5	3	3	2	3	3
4	3	3	3	3	3	2	3	4	3	3	4	3	3	3
3	3	4	4	4	4	3	4	4	3	5	4	3	2	3
3	3	3	4	3	3	3	3	5	3	4	4	3	2	4
4	4	3	5	4	4	5	3	4	4	3	5	2	4	4
4	2	2	3	3	2	4	4	4	4	3	4	4	5	3
5	5	3	3	2	3	4	5	5	4	4	4	3	4	4
4	4	2	3	4	3	2	2	3	3	3	3	2	4	4
4	3	2	3	3	3	3	3	4	3	4	4	4	3	3
2	3	3	3	3	3	4	2	4	3	4	4	2	3	3
2	3	3	4	3	3	3	4	4	3	2	4	4	2	3
1	4	4	2	3	3	2	3	2	2	2	4	5	3	3
4	3	5	3	3	3	4	5	4	3	3	4	3	3	3
3	4	5	3	3	3	3	3	4	4	2	4	4	3	3
3	4	3	3	3	3	3	4	4	4	4	4	4	4	3
2	3	3	2	2	2	4	5	4	3	3	4	4	2	3
4	2	4	3	3	3	3	3	5	3	3	4	3	4	4
2	3	3	4	3	3	3	4	4	3	4	3	2	4	3
3	3	3	3	3	4	4	5	5	5	5	4	3	2	3
2	3	2	3	3	2	3	5	5	4	5	5	1	5	4
2	4	3	3	3	3	4	3	5	5	5	4	4	5	4

TABLE 3.10 Common Software Cost Data Set.

t01 customer participation	
t02 development environment	
t03 staff availability		
t04 level and use of standards	
t05 level and use of tools	
t06 logical complexity of the software	
t07 requirements volatility	
t08 quality requirements	
t09 efficiency requirements	
t10 installation requirements	
t11 analysis skills of staff	
t12 applications experience of staff	
t13 tools skills of staff		
t14 project and team skills of staff	
t15 unadjusted experience function points	
scale: 0 (none) to 5 (considerable)	

TABLE 3.11 Pivot Table for Table 3.10.

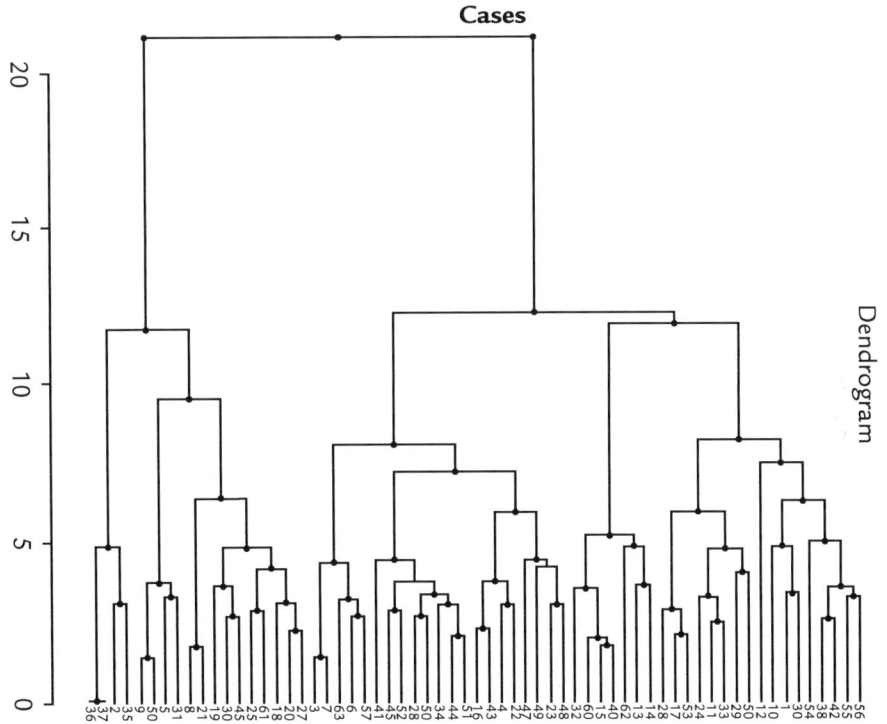

FIGURE 3.15 Common Software Cost Data Dendrogram.

Ward's method generates the best interpretation for the final clustering:

{{36, 37, 2, 35, 9, 58, 5, 31, 8, 21, 19, 39, 45, 25, 61, 18, 20, 27}, {3, 7, 63, 6, 57, 41, 46, 52, 26, 59, 34, 44, 51, 16, 43, 4, 22, 47, 49, 23, 48}, {32, 60, 15, 40, 62, 13, 14, 28, 17, 53, 24, 11, 33, 29, 50, 12, 10, 1, 30, 54, 38, 42, 55, 56}} or

{{2, 5, 8, 9, 18, 19, 20, 21, 25, 27, 31, 35, 36, 37, 39, 45, 58, 61}, {3, 4, 6, 7, 16, 22, 23, 26, 34, 41, 43, 44, 46, 47, 48, 49, 51, 52, 57, 59, 63}, {1, 10, 11, 12, 13, 14, 15, 17, 24, 28, 29, 30, 32, 33, 38, 40, 42, 50, 53, 54, 55, 56, 60, 62}}

Group I:{2, 5, 8, 9, 18, 19, 20, 21, 25, 27, 31, 35, 36, 37, 39, 45, 58, 61} has centroid:

3.28   2.61   3.61   3.11   3.33   2.78   3.83   4.44   4.50   4.22
4.06   4.22   3.39   2.56   3.78

Group II:{3, 4, 6, 7, 16, 22, 23, 26, 34, 41, 43, 44, 46, 47, 48, 49, 51, 52, 57, 59, 63} has centroid:

3.33   3.24   2.76   3.33   3.05   3.10   3.24   3.05   4.33   3.71
3.57   3.81   3.05   3.62   3.43

Group III:{1, 10, 11, 12, 13, 14, 15, 17, 24, 28, 29, 30, 32, 33, 38, 40, 42, 50, 53, 54, 55, 56, 60, 62} has centroid:

2.71   3.21   2.83   3.08   2.83   2.88   2.83   4.00   3.46   3.04
2.75   3.54   2.83   3.54   2.92

Group I stated considerable importance to quality requirements, efficiency requirements, installation requirements, analysis and application experience of staff. Group I places an emphasis on all requirements coupled with the analysis skills and application experience of the staff.

Group II stated considerable importance with respect to only installation requirements and modest concern for efficiency requirements, analysis and application experience of staff, and project and team skills of staff. Group II places an emphasis primarily on installation requirements with modest concern regarding efficiency requirements plus staff project and team as well as analysis skills combined with the staff's application experience.

Group III stated considerable importance with respect to quality requirements plus modest concern for application experience skills of staff plus project and team skills of staff. Group III places emphasis primarily on quality requirements and has a modest concern on all other issues.

3.7 EXERCISES

1. Given the ordinal proximity matrix for $n = 5$:

	x_1	x_2	x_3	x_4	x_5
x_1	0	6	2	8	7
x_2	6	0	1	5	3
x_3	8	1	0	10	9
x_4	2	5	10	0	4
x_5	7	3	9	4	0

TABLE 3.12 Hypothetical Proximity Matrix.

(a)   Generate an agglomerative hierarchical clustering derived from a sequence of threshold graphs.

(b)   Generate a divisive hierarchical clustering derived from a sequence of threshold graphs.

2. How should ties in proximity be handled for:

(a) the single-link method.

(b) the complete-link method.

3. Obtain the clustering found by applying Ward's method to the proximity matrix in problem 1.

4. Why do snake-like chain of clusters happen in single linkage methods?

5. What advantage is there for complete-link clustering as compared to single-link clustering?

6. Can the following data file have Ward's method applied to obtain a hierarchical clustering?

Defend your answer.

Runner ID	Gender	Age	Height	Weight	5k Time
1	Male	19	5' 9"	150#	17:41
2	Female	21	5' 7"	131#	21:45
3	Female	32	5' 4"	115#	23:22
4	Gender	Age	Height	Weight	Time
5	Male	38	5' 10"	185#	26:00

TABLE 3.13 Hypothetical Runner Data Set.

7. Run all the hierarchical methods discussed in the chapter on the problem 1 data set.

8. Characterize the single-link and complete-link clustering methods geometrically.

9. Use a hierarchical clustering of distances in miles between some Texas cities to obtain both a single-linkage and a complete-linkage solution. Be sure to interpret your results.

10. Use the Neymann-Scott cluster generator, presented in Jain[9] on pages 273-274, to generate an initial configuration. An important Pascal program, an implementation of the modified Neymann-Scott routine, can be used to generate samples of spherically shaped Gaussian clusters which are located randomly in the sampling window. The input parameters necessary for the Neymann-Scott algorithm include:

n, the number of points to be generated,

d, the dimensionality of space,

s, the cluster spread,

nmin, the minimum number of points per cluster,

I, the overlap index, and

c, the number of clusters desired.

The points in the d-dimensional space are independent samples from d-dimensional Guassian distributions that are centered at a randomly chosen cluster center. The overlap index, I, has a value between 0 and 1. Zero indicates well-separated clusters, and a one indicates coincident clusters. The Neymann-Scott algorithm is described by Algorithm. The IMSL normal random number generator, which generates standard normal values was used to generate the means and the points around each mean in the Neymann-Scott program.

1. Establish cluster sizes $\{n_1, n_2, \ldots, n_c\}$ for which the sum of the cluster sizes is equal to the total number of data units desired, and no cluster has less than nmin points. This is accomplished by setting n_k to nmin for all k, then selecting clusters at random and incrementing their size by 1 until the sum of the cluster sizes is n.

2. Generate cluster center μ_i at random in the sampling window (unit hypercube).

3. Scatter n_i patterns around μ_i according to a $N(\mu_i, \sigma^2)$ distribution. Reject patterns falling outside the sampling window. Continue until n_i patterns have been generated inside the sampling window.

4. If any of the overlaps, $O(i, i-1)$, $O(i, i-2), \ldots, O(i, 1)$ exceeds I, repeat steps 2 and 3.

5. If 50 repetitions do not succeed in generating a new cluster center, increase I to the smallest overlap encountered in the 50 repetitions and repeat steps 2 to 4.

6. Repeat steps 2 to 5 for i from 1 to c.

Source: Jain and Dubes[9]

Algorithm: Neymann-Scott Algorithm for Generating Clustered Data

Use a software clustering/statistical package to generate hierarchical clusterings for a variable number of patterns and variable number of clusters. The Neymann-Scott algorithm enables researchers in cluster analysis to perform Monte Carlo studies. A Monte Carlo study involves artificial data that is generated through the use of a random number generator and the probability distribution of interest. The interested reader should study Shannon.[10]

---

[9] Jain, A. K. and Dubes, R. C. (1988). *Algorithms for Clustering Data.* Englewood Cliffs, NJ: Prentice-Hall, Inc.

[10] Shannon, R. (1975). *System Simulation The Art and Science.* Englewood Cliffs, NJ: Prentice-Hall, Inc.

PARTITION CLUSTERING

4.1 INTRODUCTION

The basic method used by the optimization techniques is to obtain an initial partition of the data units and then alter the cluster membership to get a better partition, i.e., until a local optimum is found. The various methods differ as to how they arrive at an initial partition and how they define "better" (i.e., the objective function). The techniques in this section differ from the hierarchical techniques in two ways. First, members are allowed to change cluster membership. So, a data point poorly classified in an early cycle can be reclassified in a later step. Second, the number of clusters must be specified in advance. If the structure of the data is known ahead of time, then this is not a problem. Otherwise, two possible suggestions are to: (1) try a number of different partitions, or (2) use a hierarchical technique first to help decide on the number of clusters.

4.2 ITERATIVE PARTITION CLUSTERING METHOD

The following discussion provides a strategy for grouping objects into clusters, the minimization of square-error. Essentially, the task is to find the partition for a fixed number of clusters that minimizes the square error. The reader is asked to defend the fact that Ward's method employs square-error in a distinct manner (refer to 3.4 of Chapter 3). Such a strategy is called a method, not an algorithm. An algorithm is the development or foundation leading to a computer program which implements the strategy.

Iterative Partition Clustering Method

Step 1: Start with the fixed number of clusters and select an initial partition of the objects.

Step 2: After determining the cluster centers, assign each object to the object's nearest cluster center.

Step 3: Determine the new cluster centers, or centroids of the clusters, based on the new partition created by the completion of step 2.

Step 4: Repeat steps 2 and 3 until an optimum value of the objective function is achieved.

Step 5: If possible, adjust the number of clusters through merging and splitting existing clusters. At this time, removal of cluster outliers can be made.

Repeat Steps 2–5 until cluster membership stabilizes.

This is a computationally explosive problem. Some authors like Dubes and Jain[1] state that there are only 34,105 distinct partitions of 10 objects into 4 clusters, which explodes into 11,259,666,000 for 19 clusters. How

---

[1] Jain, A. K. & Dubes, R. C. (1988). *Algorithms for Clustering Data.* Englewood Cliffs, NJ: Prentice Hall.

would the researcher overcome this road block? Manual computations, or even programmed efforts on a small computer, are not the answer.

The use of an objective function is an answer. Objective functions are intentionally restricted to be evaluated for a small set of reasonable partitions. The hill climbing technique is particularly used, which is a localized greedy algorithmic approach and often found in artificial intelligence applications. These algorithms tend to be computationally efficient but sometimes converge to local minima of the objective function.

The following example illustrates the computations involved in an iterative partition method. Suppose the following partition of points in 2-dimensional Euclidean space and the initial centroids, chosen randomly, are given by:

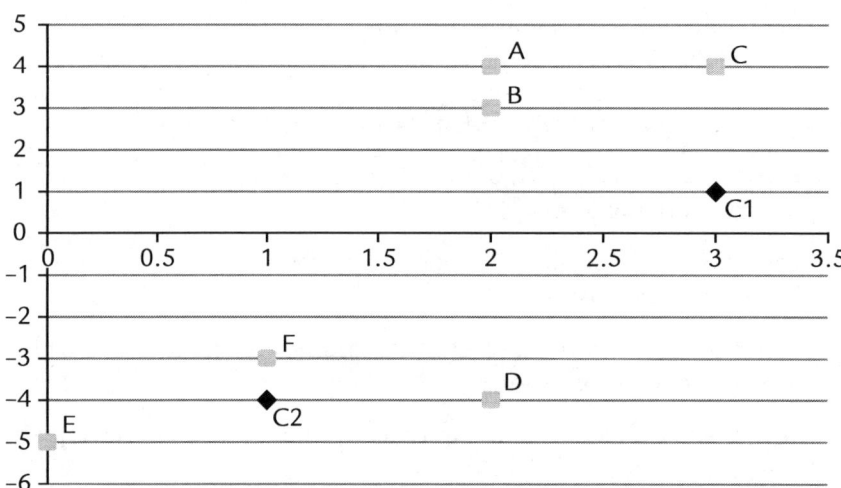

FIGURE 4.1 Randomly Generated Data.

Then the distance matrix generated, using the Euclidean distance metric, is:

C1	C2	A	B	C	D	E	F
O	5.39	3.16	2.00	3.00	5.10	6.71	4.47
5.39	0	8.06	7.07	8.25	1	1.41	1

TABLE 4.1 Euclidean Distance Matrix.

The first row of the distance matrix contains the distance of each point from C1, the first centroid. The second row contains the distances of each point from C2, the second centroid. For instance, the distance of C from C1 is $\sqrt{(3-3)^2 + (1-4)^2}$ and from C2 is $\sqrt{(1-3)^2 + (4+4)^2}$. The resultant membership matrix, where a value of 1 indicates the object is assigned to the cluster while 0 indicates nonmembership, generated is:

C1	C2	A	B	C	D	E	F
O	1	1	1	0	0	0	0
1	0	0	0	1	1	1	1

TABLE 4.2 Membership Matrix.

Because the memberships have been resolved, we can compute the new centroids. Note that each centroid is simply the average coordinate among the members of the group. Thus,

$$C1 = \left(\frac{2+2+3}{3}, \frac{4+3+3}{3} \right) = (2.33,\ 3.66) \text{ and}$$

$$C2 = \left(\frac{2+0+1}{3}, \frac{(-4)+(-5)+(-3)}{3} \right) = (1,\ -4)$$

resulting in:

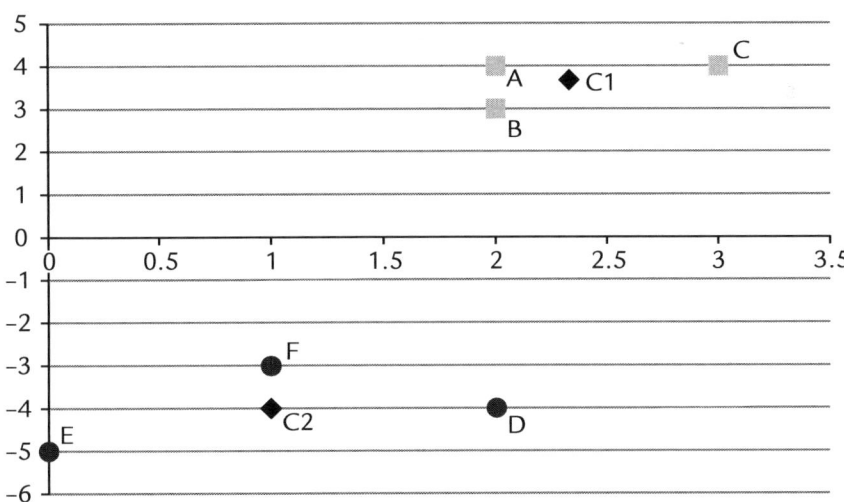

FIGURE 4.2 First Iteration Centroids.

At the end of the first iteration, C1 has been located closer to the actual cluster of {A, B, C} while C2 is still close to a potential group of {D, E, F}. None of the data points change cluster membership; therefore, this completes the overall process.

4.3 THE INITIAL PARTITION

Once the number of clusters has been decided on, the next step is to decide on the initial partition (i.e., to which clusters the data points will initially be assigned). There are a number of different methods that can be used. In the following discussion, let k refer to the number of clusters, and n_i refer to the number of entities in the ith cluster ($i = 1$ to k: $n_1 + n_2 + \ldots + n_k = n$).

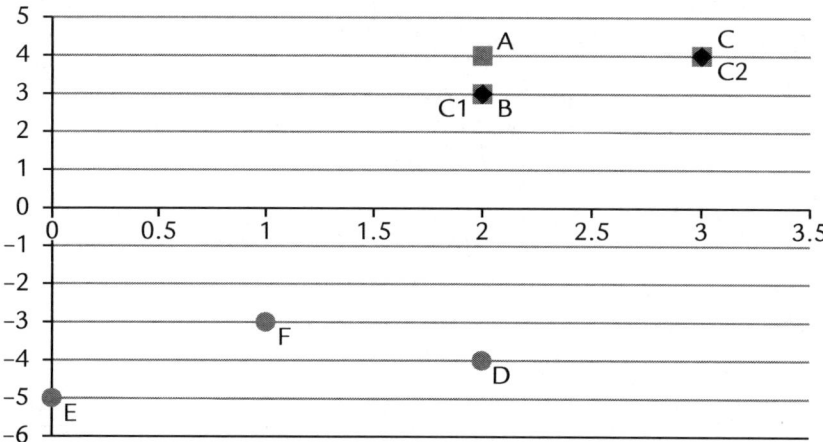

FIGURE 4.3 2nd Iteration Data Set.

Method 1: The simplest method is to choose the first k data points to act as cluster centers.

C1	C2	A	B	C	D	E	F
O	5.52	0.5	0.5	1.11	7.5	10.2	6.58
5.52	0	6.02	5.02	6.02	4.03	7.15	5.02

TABLE 4.3 2nd Iteration Distance Matrix.

C1	C2	A	B	C	D	E	F
O	1	1	1	1	0	0	0
1	0	0	0	0	1	1	1

TABLE 4.4 2nd Iteration Membership Matrix.

In the last example, set the clusters to {A, B, C} and {D, E, F}. Using method 1 results in C1 = B and C2 = C:

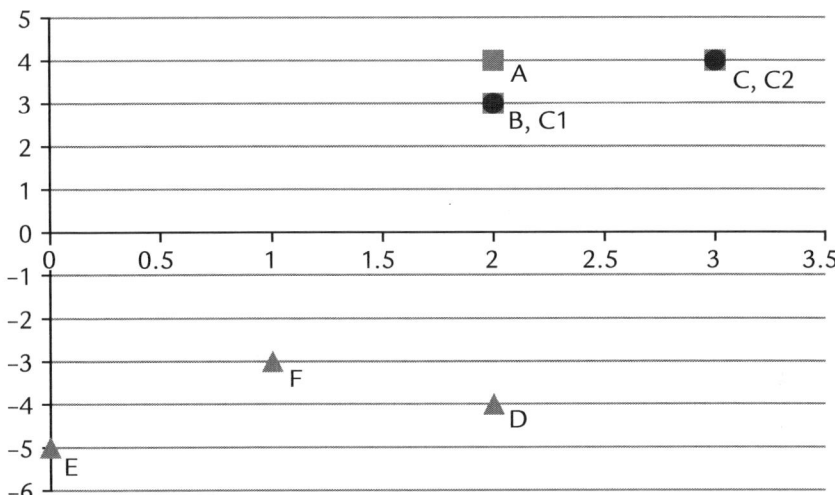

FIGURE 4.4 First *k* Data Points Used for Initial Partition.

The following is the entry for the data:

2	3	B
3	4	C
1	–3	F
2	4	A
0	–5	E
2	–4	D

The distance matrix becomes the data shown in Table 4.5.

C1	C2	A	B	C	D	E	F
O	1.41	1	0	1.41	7.0	8.25	6.08
1.41	0	1.41	1.41	0	9.64	9.48	7.28

TABLE 4.5 Initial Partition Distance Matrix.

The associated membership matrix is shown in Table 4.6.

C1	C2	A	B	C	D	E	F
1	0	1	1	0	1	1	1
0	1	0	0	1	0	0	0

TABLE 4.6 Initial Partition Membership Matrix.

New centroids [C1 = (3.4, –1) and C2 = C], generating a new partition C, D, E, and F where C has changed its cluster membership, is shown in Figure 4.5.

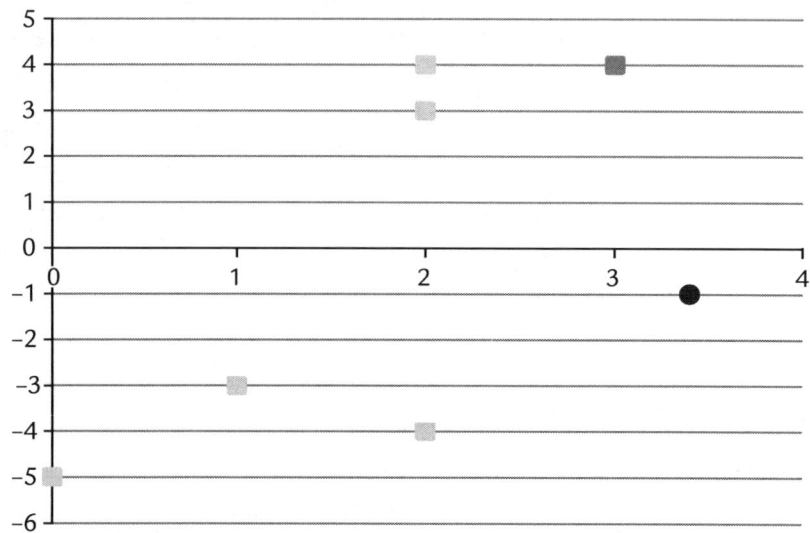

FIGURE 4.5 First Iteration Data and Centroids.

At the end of this iteration, the clusters are {A, B, D, E, F} and {C}. This radical change in cluster membership demonstrates the impact that the initial partitioning has upon the results generated in an iterative partition method.

Method 2: A second method that tries to account for any nonrandom sequence is to number the entities from 1 to n and choose the integer parts of n/k, $2n/k$, . . ., $(k - 1)n/k$ and n for the initial cluster centers.

For the current example, $n = 6$ and $k = 2$. Then we would choose points 6/2 and (2*6)/2 or points C1 = (3, 4) and D = (2, –4). In this case, we would retain the original clusters at the end of the first iteration.

Method 3: A third method is to choose the k data points randomly from the data set. In these three methods, once the k seed points have been chosen, the remaining entities are then placed in the cluster that they are closest to, usually using the Euclidean squared distance.

Method 4: A fourth method is to partition the data units so that the first n_1 in the first partition, the next n_2 are in the second cluster, and the last n_k are in the k cluster.

The methods listed previously, as well as others, are described in more detail in Anderberg.[2]

Method 5: A final method, which is discussed by Hartigan[3] is to decide on the appropriate cluster for each individual data unit by use of the formula:

$$C = K \left| \sum_{i=1}^{p} V_i \right| / (\text{Max} - \text{Min})$$

where, C is the cluster number to assign to the data unit (C = 1, 2,..., K); K is the total number of clusters; V_i is the number of variables being used to cluster (i =1, 2,..., p); Min is the minimum score when all the variables arc summed for each data unit; and Max is the maximum score when all the variables are summed for each data unit.

4.4 THE SEARCH FOR POOR FITS

Once the data points have been allocated to an initial partition, the next step is to search for poor fits. The two most popular types of clustering criteria are variations on the k-means technique and procedures based on the fundamental matrix equation

$$T = W + B$$

where T is the total dispersion matrix and W and B are the matrices of within-cluster and between-cluster variation, or scatter.

Assume the data points being studied have d-dimensional patterns and they are separated into K clusters, n_k rows for cluster k, k = 1 to K. Then the data columns for the patterns in the kth cluster are $\left[\vec{x}_1^{*(k)}, \vec{x}_2^{*(k)}, \ldots, \vec{x}_{n_k}^{*(k)} \right]^T$ where $\vec{x}_j^{*(k)} = \left[\vec{x}_{j1}^{*(k)}, \vec{x}_{j2}^{*(k)}, \ldots, \vec{x}_{jd}^{*(k)} \right]^T$. For the ith feature in the kth cluster, the mean is:

$$m_i^{(k)} = \left(\frac{1}{n_k} \right) \sum_{j=1}^{n_k} x_{ji}^{(k)}$$

Then the vector of feature means for the kth group is:

$$\vec{m}^{(k)} = \left[m_1^{(k)}, m_2^{(k)}, \ldots, m_d^{(k)} \right]^T$$

[2] Anderberg, M. R. (1973). *Cluster analysis for applications.* New York: Academic Press.
[3] Hartigan, J. A. (1975). *Clustering Algorithms.* New York: John Wiley & Sons.

Therefore the grand mean, also a vector, is:

$$\vec{m} = \left(\frac{1}{n}\right)\sum_{k=1}^{K} n_k \vec{m}^{(k)} \text{ where } n = \sum_{k=1}^{K} n_k$$

Let $x_j^{(k)} = x_j^{*(k)} - \vec{m}$

Then $T = \sum_{k=1}^{K}\sum_{j=1}^{n_k} \left(\vec{x}_j^{(k)}\right)\left(\vec{x}_j^{(k)}\right)^{T}$ is the total scatter matrix.

The within scatter matrix for the kth cluster is:

$$W^{(k)} = \sum_{j=1}^{n_k} \left(\vec{x}_j^{*(k)} - \vec{m}^{(k)}\right)\left(\vec{x}_j^{*(k)} - \vec{m}^{(k)}\right)^{T}$$

and the total within scatter matrix is:

$$W = \sum_{k=1}^{K} W^{(k)}$$

Then B, the between scatter matrix is:

$$B = \sum_{k=1}^{K}\sum_{j=1}^{n_k} (\vec{m}^k - \vec{m})(\vec{m}^k - \vec{m})^{T}$$

K-means clustering is to minimize the clustering objective function.

$$min_H J_K, \text{ where } J_K = \sum_k \sum_{i \in C_k} \left\|x_i - m_k\right\|^2$$

and $H = \{0, 1\}^{n \times K}$ where $H_{jk} = 1$ *if* $x_j \in C_k$ and 0 otherwise.

It is well known that T = W + B. In the case of k-means clustering we have:

$$J_K = Trace\,(W) = Trace\,(T - B)$$

Note that Trace(T) is a constant. Therefore k-means minimizes the within-cluster scatter matrix or maximizes the between-cluster scatter matrix.

The following discussion illustrates the above formalism with a numerical example. Given the data points

$$X = \begin{bmatrix} 0 & -5 \\ 1 & -3 \\ 2 & -4 \\ 2 & 4 \\ 2 & 3 \\ 3 & 4 \\ 3 & 3 \end{bmatrix} \text{ and } X^{T} = \begin{bmatrix} 0 & 1 & 2 & 2 & 2 & 3 & 3 \\ -5 & -3 & -4 & 4 & 3 & 4 & 3 \end{bmatrix}$$

where the first three rows of X are members of cluster 1 and the last four rows are members of cluster 2.

$$m_1^{(1)}=\left(\frac{1}{3}\right)(0+1+2);\ m_2^{(1)}=\left(\frac{1}{3}\right)((-5)+(-3)+(-4))\ \text{or}\ \vec{m}^{(1)}=\begin{pmatrix}1\\-4\end{pmatrix}.$$

In a similar fashion, $\vec{m}^{(2)}=\begin{pmatrix}2.5\\3.5\end{pmatrix}$. Then

$$\vec{m}=\left(\frac{1}{n}\right)\left[(3)\begin{pmatrix}1\\-4\end{pmatrix}+(4)\begin{pmatrix}2.5\\3.5\end{pmatrix}\right]\ \text{or}\ \vec{m}=\begin{pmatrix}1.86\\0.29\end{pmatrix}.\ \text{Next find T:}$$

$$\vec{x}_1^{(1)}=\begin{pmatrix}1\\-5\end{pmatrix}-\begin{pmatrix}1.86\\0.29\end{pmatrix}=\begin{pmatrix}-1.86\\-5.29\end{pmatrix}$$

$$\vec{x}_2^{(1)}=\begin{pmatrix}-0.86\\-3.29\end{pmatrix}$$

$$\vec{x}_3^{(1)}=\begin{pmatrix}0.14\\-4.29\end{pmatrix}$$

$$\vec{x}_1^{(2)}=\begin{pmatrix}0.14\\3.71\end{pmatrix}$$

$$\vec{x}_2^{(2)}=\begin{pmatrix}0.14\\2.71\end{pmatrix}$$

$$\vec{x}_3^{(2)}=\begin{pmatrix}1.14\\3.71\end{pmatrix}$$

$$\vec{x}_4^{(2)}=\begin{pmatrix}1.14\\2.71\end{pmatrix}$$

Calculate the following:

$$\vec{x}_1^{(1)}\vec{x}_1^{(1)^T}=\begin{pmatrix}-1.86\\-5.29\end{pmatrix}(-1.86-5.29)=(-1.86)^2+(-5.29)^2=31.44$$

$$\vec{x}_2^{(1)}\vec{x}_2^{(1)^T}=11.56$$

$$\vec{x}_3^{(1)}\vec{x}_3^{(1)^T}=18.42$$

and $T^{(1)} = 61.42$.

The computation for $T^{(2)} = 45.24$ is left to the reader. Thus $T = 106.66$.

Next, find W:

$$\left[\vec{x}_1^{(1)} - \vec{m}^{(1)}\right]\left[\vec{x}_1^{(1)} - \vec{m}^{(1)}\right]^T = \left[\begin{pmatrix} 0 \\ -5 \end{pmatrix} - \begin{pmatrix} 1 \\ -4 \end{pmatrix}\right]\left[\begin{pmatrix} 0 \\ -5 \end{pmatrix} - \begin{pmatrix} 1 \\ -4 \end{pmatrix}\right]^T = \begin{pmatrix} -1 \\ -1 \end{pmatrix}(-1 \quad -1) = 2$$

with similar computations yields: $W^{(1)} = 4$ and $W^{(2)} = 2$, or $W = 6$. Finish by deriving B:

$$\vec{m}^{(1)} - \vec{m} = \begin{pmatrix} 1 \\ -4 \end{pmatrix} - \begin{pmatrix} 1.86 \\ 0.29 \end{pmatrix} = \begin{pmatrix} -0.86 \\ -4.29 \end{pmatrix} \text{ and } \left[\vec{m}^{(1)} - \vec{m}\right]\left[\vec{m}^{(1)} - \vec{m}\right]^T = 19.14$$

and with comparable computations $\left[\vec{m}^{(2)} - \vec{m}\right]\left[\vec{m}^{(2)} - \vec{m}\right]^T = 10.71$. Then $B = (3)(19.14) + (4)(10.71) = 100.27$. Note that $T = 106.66 = W + B$, allowing for rounding errors. Now a Jancey k-means clustering can be completed, see problem 4 and 5 in the exercises found at the end of the chapter.

4.5 K-MEANS ALGORITHM

In the k-means approach, once the initial partitions have been established, the cluster centroid can be updated after each addition to the cluster (MacQueen's method), or, only after all the points have been allocated (Forgy's method and Jancey's method).

At this point a local optimum has been achieved. Unfortunately, there is no way to know whether this is also the global optimum. The solution can either be accepted, or a new initial partition can be obtained and the clusters can be reanalyzed.

4.5.1 MacQueen's Method

The basic algorithm for MacQueen's method is as follows:

1. Begin with an initial partition of the data units into clusters (any of the methods mentioned previously will work).

2. Take each data unit in sequence and compute the distances to all cluster centroids; if the nearest centroid is not that of the data unit's parent cluster, then reassign the data unit and update the centroids of the losing and gaining clusters.

3. Repeat step 2 until convergence is achieved; that is, continue until a full cycle through the data set fails to cause any changes in cluster membership (Anderberg[1], p. 162).

4.5.2 Forgy's Method

The basic algorithm for Forgy's method is to:

1. Begin with any desired initial configuration. Go to step 2 if beginning with a set of seed points; go to step 3 if beginning with a partition of the data units.

2. Allocate each data unit to the cluster with the nearest seed point. The seed points remain fixed for a full cycle through the entire data set.

3. Compute new seed points as the centroids of the clusters of data units.

4. Alternate steps 2 and 3 until the process converges; that is, continue until no data units change their cluster membership at step 2 (Anderberg[1], p. 161).

4.5.3 Jancey's Method

Though Jancey's method is one of the clustering methods falling under the k-means label, it is interesting to note that his method implicitly minimizes a within cluster error function, and as such, may be regarded as attempting to minimize the trace of **W** (Anderberg)[1];(Everitt)[4]; and (Seber).[5] Jancey's procedure begins by choosing k mutually exclusive groups and computing the group centroids, which he calls class points. In the present investigation, the method attributed to Hartigan was used to develop the initial partitions. The next step is to search for objects that should be reallocated to another cluster. In Jancey's method, this is accomplished by allocating each data unit to the cluster with the nearest class point. The class points remain fixed for a full cycle through the entire data set. At all succeeding stages, each new class point is found by reflecting the old class point through the new centroid for the cluster. Repeated passes are made until no further improvement can be made by moving an object, i.e., the process converges. The interested reader should refer to Jancey[6] for his reason for reflection.

---

[4] Everitt, B. S. (1980). *Cluster analysis (2nd ed.)*. New York: Halsted Press.
[5] Seber, G. A. F. (1984). *Multivariate observations.* New York: John Wiley and Sons.
[6] Romesburg, H. C. (1984). *Cluster Analysis for Researchers.* Belmont, CA: Wadsworth, Inc.

Consider writing a program for clustering data according to Jancey's k-mean algorithm. This program should allow the user to specify one of three initial configuration methods; none of the methods require information from the user. The initial configuration techniques employed are to be selected in order to demonstrate the benefit of choosing the initial seeds by mathematical methods instead of relying on the user to select the seeds. In the first method, which is analogous to the user selecting k points from the data set, k distinct data points are randomly chosen as initial seeds through the use of a random number generator. A multiplicative congruent random number generator was implemented. The pseudocode for this random number generator is described in Algorithm 1.

---

1. Choose any number less than nine digits and designate it as XQ, the starting value. This seed should be chosen randomly, perhaps by using a table of random digits.

2. Multiply this by a designated number, a, a nonnegative number of at least five digits.

3. Multiply the product from step 2 by a fraction or decimal number that is equal to $1/m$, where m is a nonnegative integer equal to $2**b$ and b is the number of bits in a computer word.

4. Choose the decimal portion of the answer from step 3 as a random number $0 < X < 1$.

5. Drop the decimal point from the original number obtained in step 4 and use it for the X to be multiplied by a in step 2.

6. Repeat steps 2 through 5 until the desired number of random numbers is obtained.

FIGURE 4.6 Algorithm 1: Multiplicative Random Number Generator.

Secondly, all data points are ordered according to the sum of the numerical attributes which describe each point. Then, the ordered data points are divided into k groups. The means of these k groups are then calculated and serve as the initial seeds. Finally, the SAS FASTCLUS routine generated the third initial configuration. The SAS FASTCLUS program accepts as input parameters the initial seeds computed in the second initial configuration method; it then generates a refined set of seeds. The algorithm used by SAS FASTCLUS to calculate the refined initial seeds is contained in Algorithm 2.

1. It accepts as input parameters k seeds which have a large Euclidean distance between them. The k seeds cover all regions of the attribute space in which the objects are expected to reside.

2. It then forms temporary clusters by sequentially assigning the remaining objects to the cluster seed each object is nearest. As objects are assigned, the cluster seed is recomputed and made to be equal to the mean of the data profiles of the objects that are in the cluster. As a consequence, the cluster seed (usually) changes when an object is tentatively assigned to a cluster.

3. When the first iteration is completed, the final set of cluster seeds are taken as the k initial seeds to start the second iteration. The process is repeated, sequentially assigning the objects to their nearest cluster seed, and updating the seeds as the process moves along.

4. After a number of iterations, when the change in the positions of the cluster seeds is tolerably small from one iteration to the next, the program terminates. The k cluster seeds from the final iteration serve as the refined seeds.

Source: Romesburg[6]

FIGURE 4.7 Algorithm 2: SAS FASTCLUS Algorithm for Generating Initial Seeds.

Jancey's algorithm for clustering the data points is displayed in Algorithm 3. After clustering the data according to Jancey's algorithm, the program should calculate the Corrected Rand Statistic, a statistical index measuring the departure of the captured partition from the actual partition.

1. Begin with any desired initial configuration.

2. Allocate each data unit to the cluster with the nearest seed point. The seed points remain fixed for a full cycle through the entire data set.

3. Compute new seed points as the centroids of the clusters of the data units.

4. Iterate steps 2 and 3 until the process converges; that is, continue until fewer than the tolerable number of data units change their membership at step 2.

Source: Anderberg[2]

FIGURE 4.8 Algorithm 3: Jancey's Algorithm for Clustering Data.

The Corrected Rand Statistic is an external measure. External measures are usually calculated based on how pairs of points are placed in the actual and captured clusters. There are four possibilities of placement for a pair of points. First, the pair is placed in the same cluster in the actual and captured partitions. Secondly, the pair is placed in the same cluster in the actual partition and in different clusters in the captured partition. Alternatively, the pair is placed in different clusters in the actual partition and in the same cluster in the captured partition. Finally, the pair is placed in different clusters in the actual and captured partitions. The first and last cases represent similarity between partitioning; the middle two placement schemes represent differing partitions. Many of the external measuring indices are calculated by summing up the number of pairs of points which fall into each of the four categories mentioned previously. Then the actual index is a function of these four terms. For example, the Rand Statistic is calculated by summing the first and last term; and then dividing this sum by the sum of all the terms. The Corrected Rand values must fall between zero and one. This statistic is one of the newest external measures in the cluster analysis field. It is based on the Rand Statistic. Hubert and Arabic introduced the Corrected Rand Statistic; this statistic corrects the original rand index for chance. The algorithm for computation of the Corrected Rand Statistic is contained in Figure 4.9.

---

1. Calculate the index term which equals the $\sum_{i,j} (n_{ij}\ choose\ 2)$. Note that $i = 1$ to the number of clusters in the actual partition, and $j = 1$ to the number of clusters in the captured partition, n_{ij} = number of data points in cluster i of the actual partition and cluster j of the captured partition.

2. Calculate the expected index term. To do this calculate

$$\left[\sum_i (ni\ choose\ 2) * \sum_j (nj\ choose\ 2)\right] \Big/ [n\ choose\ 2]$$

Note that n_i = number of data points in cluster i of the actual partition and n_j = number of data points in cluster j of the captured partition, n = the total number of data points clustered.

3. Calculate the maximum index term. To do this calculate

$$\left[\sum_i (ni\ choose\ 2) + \sum_j (nj\ choose\ 2)\right] \Big/ 2$$

4. Calculate the Corrected Rand Statistic which is equal to (Index term – Expected Index term) / (Maximum Index term – Expected Index term).

Source: Hubert and Phipps[8]

FIGURE 4.9 Algorithm 4: Algorithm for Computing the Corrected Rand Statistic.

4.6 GROUPING CRITERIA

Looking at the matrix relationship mentioned earlier (i.e., W = T + B), four possible grouping criteria have been proposed.

Criteria 1: The first possibility is to minimize the trace W. This is one of the three most popular techniques of cluster analysis (Everitt).[3]

Criteria 2: Second, minimize the ratio of determinants $\left|\dfrac{W}{T}\right|$. This criterion is known as Wilks' lambda statistic.

Criteria 3: Third, maximize the largest eigenvalue of W. This criterion is known as Roy's largest root criterion.

Criteria 4: Fourth is to maximize the trace of W^{-1} B. This criterion is known as Hotelling's trace criterion (Anderberg).[1]

Each of these methods has its advantages and disadvantages. The major problem with minimizing the trace W is that it is transformation dependent. This means, that in general, different results will be obtained from applying the technique to raw data as opposed to standardized data (Everitt).[3] A second problem with this criterion is that it tends to find spherical clusters in the data. This is because the trace does not take into account correlations among the variables (Seber).[4] The second method, minimize the ratio of determinants $\dfrac{|W|}{|T|}$, assumes that all the clusters in the data have the same shape. It also tends to divide the data into approximately equal-sized clusters when the separation is not great. Similar problems can be found for the other two criterion see Everitt[3] and Seber.[4]

4.7 BIRCH, A HYBRID METHOD

There is a need for more efficient clustering methods for large databases. Zhang, Ramakrishnan, and Livny[7] developed BIRCH, Balanced Iterative Reducing and Clustering using Hierarchies, for clustering large numerical

---

[7] Zhang, T., Ramakrishnan, R., and Livny, M. (1996). An efficient data clustering method for very large databases. *Proceedings of ACM SIGMOD International Conference on Management of Data,* 103-114.

databases. The BIRCH method combines the construction of a height-balanced tree with a k-means method applied to the leaves of the tree.

Two major concepts in the BIRCH method are **cluster feature vectors** and the **cluster feature tree**.

Definition: a **cluster feature vector** is a triplet, (CF, LS, SS), maintained on a set of data points, $\{x_1, x_2, \ldots, x_N\}$, where N is the number of points, $LS = \sum_{i=1}^{N} x_i$, and $SS = \sum_{i=1}^{N} x_i^2$.

Given the following set of points, as a candidate cluster:

Point	X	Y
1	2	3
2	4	2
3	4	5
4	3	5
5	3	3
6	4	3

Then CF = (6, (20, 22), (70, 81)).

The cluster feature vectors represent the structure of the clusters, while the cluster feature tree stores the clustering hierarchy.

Definition: a **CF tree** is a height-balanced tree which stores the clustering features for a hierarchical clustering with two parameters: B = the maximum number of children for each parent node and L = the maximum diameter of sub-clusters stored at a leaf node. Each entry in a non-leaf node has the form $[CF_i, child_i]$, where $child_i$ is a pointer. The leaf nodes are stored in a doubly linked list, where each entry is a cluster feature vector, CF_i. All leaf nodes are on the same level of the tree.

A template for a CF tree is illustrated in Figure 4.10:

B = 5, L = 7

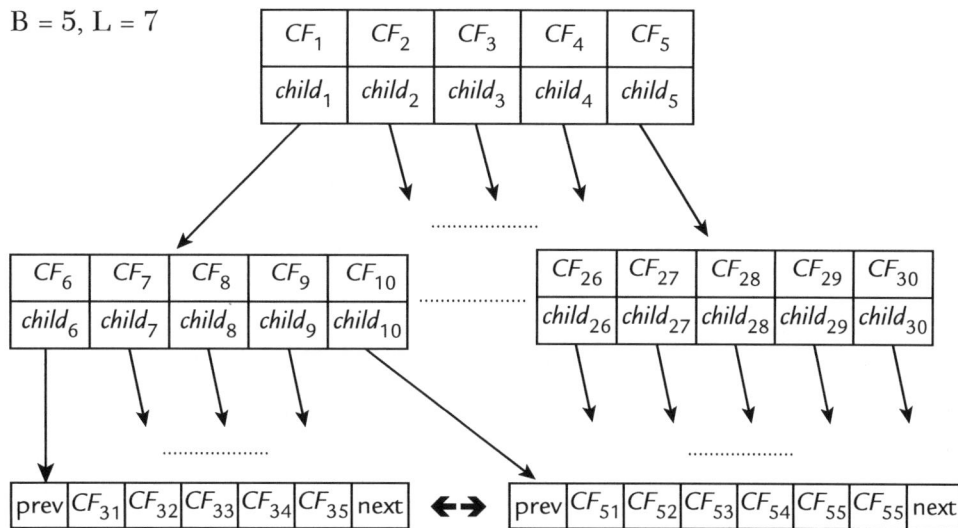

FIGURE 4.10 A CF tree.

The BIRCH algorithm is composed of two main phases:

Phase A: construction of a memory resident CF tree for the numerical database, and

Phase B: a partitioning algorithm, similar to k-means, is applied to the leaves of the CF tree.

The following Phase A pseudocode is for constructing the CF tree:

Input: the database $D = \{t_1, t_2, \ldots, t_n\}$ and T = the maximal diameter for subclusters.

Output: K = the number of captured clusters.

For each $t_i \in D$ do

{find the correct leaf node to insert t_i by a search path starting at the root node

IF the number of points in the leaf node is ≤ T,

THEN insert t_i into the leaf node and update the ancestor CF triples using summation

ELSE split the leaf node, placing t_i in one of the new leaf nodes, and update the ancestor CF triples where necessary.}

Note that insertions are made at the bottom level of the tree and can result in splitting nodes in prior levels of the tree, even allowing for the tree's height to increase by one. Once the initial CF tree is attempted to be created, and if there is insufficient memory to construct the CF tree, simply restart the construction with a larger value of T. As T increases, the size of the constructed CF tree decreases exponentially.

Phase B applies another global clustering approach applied to the leaf nodes in the CF tree, by utilizing the double-linked list of leaf nodes for navigation. This phase is followed by reclustering all points by placing them in the cluster which has the closest centroid, an optional phase.

The advantage that BIRCH has is it only requires a single scan of the database. However, additional scans to consider a variety of values for T could improve the quality of the resultant CF tree choice. Not only is BIRCH applicable for numeric data, but it is also sensitive to the input order of the data. BIRCH additionally favors clusters with spherical shape and similar sizes.

4.8 SUMMARY

- Unlike hierarchical clustering methods that capture a tree structure of clustering structure, the partition clustering methods obtain a single partition of the data. When the computational requirement for the construction of a dendrogram is prohibitive, then a partition clustering method should be used for the cluster analysis.

- Choosing the number of representative clusters is the paramount problem in partition clustering. The interested reader should refer to Dubes.[8]

- Partition clustering methods generate clusters either by optimizing an objective function on a subset of the objects or over all of the objects.

- Normally, in a cluster analysis utilizing partition clustering methods, the methods are executed multiple times with distinct initial configurations or the best configuration derived from all of the executions is adopted as the final output clustering.

- K-means, based upon the squared error criterion, is the most commonly used partition clustering method.

---

[8] Dubes, R. C. (1987). How many clusters are best?—An Experiment. *Pattern Recognition*, 20:645-663.

■ The k-means method is sensitive to the selection of the initial partition, which can cause the convergence to a local minima of the objective function.

■ The standard error for a clustering containing k clusters is:

$$\text{error} = \sum_{j=1}^{K} \sum_{i=1}^{n_j} \left\| x_i^{(j)} - c_j \right\|^2,$$

where $x_i^{(j)}$ is the ith pattern belonging to the jth cluster and c_j is the centroid of the jth cluster.

■ K-means converges but it finds a local minimum of the objective function.

■ K-means works only for numerical observations, for categorical and mixture observations k-medoids is a clustering method.

■ Fine tuning is required when applied for image processing; mostly because there is no imposed spatial coherency in k-means algorithm.

■ K-means often works as a starting point for sophisticated clustering applications.

4.9 EXERCISES

If you have access to clustering software, the following problems should be run with it instead of the software found in the Appendices.

1. How many clusters are present in the data?

x	y
1	1
2	2
5	5
4	7
5	7
5	8
4	8
14	7
15	8
19	16
19	17

TABLE 4.7 Data Set.

Defend your answer.

2. Using Forgy's method obtain a partition clustering. Discuss how the clustering in problem 1 compares to the captured clustering using Forgy's method.

3. For the problem 1 data set, obtain a partition clustering for each selection of an initial configuration discussed in section 3.3.

4. Complete the computations for the within-cluster matrix discussed in section 3.4.

5. Explain why $T = W + B$.

6. Complete a k-means clustering by Jancey's method for the pattern matrix in section 3.4.

7. Given the following data set:

Person	Weight	Height
A	135#	5'2"
B	142#	5'2"
C	175#	6'0"
D	189#	5'10"

TABLE 4.8 Physical Data.

Manually perform a partition clustering on this data given that the original set of centroids are C1(138#, 5'3") and C2(182#, 5'4"). Include for each iteration the distance matrix, the membership matrix, and a graphical display of the iteration step cluster that includes the new centroids.

8. Manually compute the corrected Rand statistic for the partition clustering obtained for problem 1 assuming the actual clusters were {A, B} and {C, D}. Interpret your result.

9. Use the Neymann-Scott cluster generator to generate an initial configuration. Redo problems 2 and 5 for this configuration.

10. Perform a cluster analysis of the data file found in problem 1 using an accessible statistical software package including cluster analysis facilities. Compare the clustering with the resultant clustering obtained from applying Jancey's algorithm to the same data file.

11. Splitting and merging of the resulting clusters can be taken as the last step in each iteration of the k-means method. A split in a given cluster can be taken if the cluster variance is above a prespecified threshold. Similarly, two clusters are merged when the distance between their centroids is below another prespecified threshold. Redo problem 7 using this modification. *Enforcing this modification to the k-means method makes it possible to obtain the optimal partition starting from any initial partition, provided proper threshold values are specified (Ball, G. H. and Hall, D. J.).*[9]

12. The Neymann-Scott algorithm enables researchers in cluster analysis to perform Monte Carlo studies. A Monte Carlo study involves artificial data that is generated through the use of a random number generator and the probability distribution of interest. The interested reader should study Shannon.[10] Consider conducting a Monte Carlo evaluation on the optimal initial configuration for the k-means methods discussed in Section 3.5.

---

[9] Ball, G. H. and Hall, D. J. (1965). ISODATA, a Novel Method of Data Analysis and Classification, Technical Report, *Stanford Research Institute*, California.

[10] Shannon, R. (1975). *System Simulation The Art and Science*. Englewood Cliffs, N. J.: Prentice-Hall, Inc.

13. Consider the following data points:

Point	x	y
1	1	1
2	2	2
3	3	3
4	4	4
5	4	5
6	4	6
7	5	5
8	6	6
9	7	7
10	8	8
11	7	6
12	6	7
13	6	8
14	8	6
15	9	10
16	3	7
17	7	3
18	6	4
19	4	6
20	4	3
21	3	2

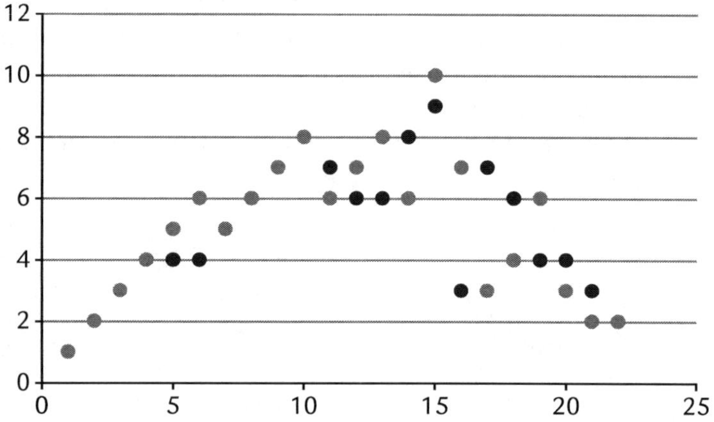

FIGURE 4.11 Hypothetical Data Set.

Construct a CF tree with B = 3 and L = 2 for the data points.

14. Use k-means to obtain a final BIRCH clustering for the data points in problem 13.

15. Implement the BIRCH algorithm in a programming language of your choice. Generate the solutions to problems 13 and 14 using your program.

JUDGMENTAL ANALYSIS

In This Chapter

5.1 INTRODUCTION

Bottenberg and Christal[1,2] developed an application of the hierarchical grouping model that groups criteria in terms of the homogeneity of their prediction equations. A specific application of this technique, JAN, was developed by Christal.[3,4] JAN is a technique for capturing and clustering the policies of raters or judges. Capturing the policy of a rater (or judge) can be defined as the extent to which one is able to predict the behavior or actions of that rater from the known characteristics of the stimuli he (she) is being required to evaluate. Clustering refers to the grouping of raters relative to the homogeneity of their prediction equations as discussed by Christal[5] and Dudycha.[6]

The following example will help illustrate the JAN procedure. A group of 10 judges are recruited to establish an evaluation policy for a new software development program. Relevant information (i.e., 12 quality assurance measures) is individually coded for a sample of 175 software assurance departments from several companies. The judges are asked to look at each department's profile and rate their suitability for the new program. A judge's policy can be captured by using the 12 quality assurance measures as predictors and the judge's rating of the 175 profiles as the criterion variable in a 12-predictor multiple regression equation. We are able to determine how successfully the regression equation predicts a given judge's policy by the magnitude of R^2, the squared multiple correlation coefficient. Once a judge's policy has been captured, with his/her individual regression equation, the problem becomes one of reaching a consensus (the clustering

[1] Bottenberg, R. A. & Crystal, R. E. (1961, March). An iterative clustering criteria which retains optimum predictive efficiency. *(Technical Note WADD - TN - 6 30)*. Lackland Air Force Base, Texas Personnel Laboratory, Wright Air Development Division.

[2] Bottenberg, R. A. & Crystal, R. E. (1968). Grouping criteria - a method which retains optimum predictive efficiency. *Journal of Experimental Education, 36*(4), 28-34.

[3] Christal, R. E. (1963, February). JAN: a technique for analyzing individual and group judgment. *(Technical Note PRL-TDR 63 - 3)*, Lackland Air Force Base, Texas 6570th Personnel Research Laboratory, Aerospace Medical Division.

[4] Christal, R. E. (1968). Selecting a harem - and other applications of the policy capturing model. *Journal of Experimental Education, 36*(4), 35-41.

[5] Christal, R. E. (1968). JAN: a technique for analyzing group judgment. *Journal of Experimental Education, 36*(4), 24-27.

[6] Dudycha, A. L. (1970). A Monte-Carlo evaluation of JAN: a technique for capturing and clustering rater policies. *Organizational Behavior and Human Performance, 5*, 501-516.

phase of JAN). First, a single R^2 is computed that gives the overall predictive efficiency obtained when a separate least-squares weighted regression equation is used for each judge. In the second and subsequent steps, all the equations are compared and the two judges (or equations) showing the most agreement (i.e., the smallest drop in the overall R^2 value) are combined to form a new equation by collapsing variables. This leaves one less equation at successive steps. The drop in R^2 indicates the overall loss in predictive efficiency that results from reducing the number of equations by one. The process continues until all the judges have been combined into one regression equation. By observing the drop in R^2 at each step of the grouping process, the number of different policies exhibited by that group of judges may be determined. There are a number of different methods for deciding upon the appropriate cutoff. Bottenberg and Christal[2] proposed an F test when the *N*s are relatively small. Ward and Hook[7] recommended looking for a sharp increase in the value-reflecting number, which "indicates that much of the classification systems accuracy has been lost by reducing the number of groups by one at this stage" (p. 73). Two methods mentioned by Adler and Kafry[8] included deciding on the number of groups a priori, or stopping the process when a predetermined lower bound of R^2 has been reached.

The previous example highlights a number of positive benefits to the JAN approach. First, it identifies the underlying groups. Second, it provides equations expressing the different policies. And third, it allows for a complete analysis of inter-rater agreement. It is not surprising that the use of this technique has become widespread since it first appeared in the journals.

5.2 JUDGMENTAL ANALYSIS ALGORITHM

In this section, definitions of various concepts related to JAN are discussed.

Ipsative JAN refers to the JAN approach in which judges use personal knowledge of the capabilities of each profile case to suggest a rank. It is a way of quantifying subjective opinions that yields more objective results.

---

[7] Ward, J. H. Jr. & Hook, M. E. (1963). Applications of a hierarchical grouping procedure to a problem of grouping profiles. *Educational and Psychological Measurement,* 23, 69-81.

[8] Adler, I. & Kafry, E. (1980). Capturing and clustering judges' policies. *Organizational Behavior and Human Performance,* 23, 384-394.

The *judges* are restricted to be knowledgeable or related to the profile cases in order to rate each profile case.

The *judgmental evaluation procedure* is the outcome of a rating process by a judge who has been presented a series of variables on which he (she) must base a decision. For example, how does a judge involve effective thought processing when presented with a series of variables on which he (she) must base a decision?

Normative JAN refers to the JAN approach in which the judges utilize the predictor variables to rank each profile case.

A *profile score* is an instance of a rating for a profile case.

Type A JAN is an analysis in which all the judges give judgments on the same subjects with respect to the same criterion variable and predictor variables.

Type B JAN is an analysis in which the judges do not judge the same subjects (different judges could judge a different number of subjects as well) with respect to the same criterion variable and predictor variables.

The judgment analysis technique is useful for identifying the rating policies existing in a board or committee of judges. With this technique, each judge makes a judgmental decision on one common criterion variable on which he (she) is given a common predictor set profile on a set of judges to determine how many different evaluation policies exist in the group. Next, a multiple regression analysis can be made to determine the composition of each captured policy.

The JAN procedure consists of two basic stages. In the first stage, a least-squares solution of a multiple regression equation is computed for each judge, which predicts the criterion decision he (she) has made. It answers the question of how consistent the judge is in his (her) use of specific variables in arriving at an overall decision by giving his (her) decision-making equation and the R^2, the square of the multiple correlation coefficient, from his (her) multiple regression analysis. A high R^2 indicates that he (she) has a policy and was consistent in expressing it through his (her) criterion decisions.

Next, a hierarchical grouping procedure is initiated that allows the grouping of individual judges according to some objective function by similarity of their decision-making equations in this case. Groupings are selected that yield the smallest decrease in ability to predict the criterion from knowledge of which individual judges are in a particular group (loss in R^2).

The first step in the grouping procedure reduces the number of equations by one. The two judges whose decision-making equations are most similar are grouped together as a unit and all other judges are left ungrouped. From the set of ungrouped judges, the process then selects a judge whose decision-making equation is most similar to that of the grouped judges and groups him (her) with them as a unit. The number of judges remaining in the ungrouped set is reduced by one at each step of the process until all judges have been grouped into a single unit. At each step of the grouping process, an examination of the loss of predictive efficiency allows the investigator to identify the number of various judgmental policies that exist.

If a single policy is characteristic of the whole collection of judges, it can be determined at the final stage of grouping. If a joint policy does not exist, arbitration of the criterion by the whole group may yield a joint policy.

Judgment analysis has been extensively used by the United States Air Force to analyze individual and group judgments and to formulate common or overall judgmental policies. A large amount of this type of work has been done by Ward[9] and Bottenberg and Christal.[1] Bottenberg and Christal[2] were originally responsible for the mathematical and computer program developments in the field.

5.2.1 Capturing R^2

Rather than solving the matrix equation developed for least squares estimators, the JAN algorithm employs an iterative technique to find the *b weights*. Such an approach is systematic, repetitive, easily programmed, and well suited to computer computation.

The Kelley-Salisbury technique is one of these methods. In it, successive approximations to β in the equation

$$R\,\beta_z = V$$

are found until a satisfactory solution is obtained. V is the vector of *validities*, or Pearson correlations between each of the predictor variables and the

---

[9] Ward, J. H., Jr., (1963). Hierarchical grouping to optimize an objective function. *American Statistical Association Journal*, 58, 236-244.

dependent variable. R is the *intercorrelation matrix* of the predictors, the matrix of correlations of the independent variables with each other.

The vector β_z found by this method is the vector of coefficients in the equation

$$z_Y = \sum_{j=1}^{p} \beta_{z_j} z_{X_{ij}}.$$

This equation is called the *normalized* solution of the multiple regression problem. We can use the facts that

$$z_{Y'} = \frac{Y' - \overline{Y'}}{s_{Y'}} \text{ and } z_{X_i} = \frac{X_i - \overline{X_i}}{s_{X_i}}, \text{ for } I = 1, 2, ..., p,$$

to find the vector b of weights in this form of the solution:

$$Y' = b_0 + \sum_{j=1}^{p} b_j X_{ij}.$$

Consider the following example, taken from a text by Walker and Lev[10] and based on a sample of $n = 36$ cases. Given the intercorrelation matrix for a dependent variable Y and the three predictors X_1, X_2, X_3, we seek the least squares multiple linear regression. The correlations of all the variables are given in Table 5.1.

	Y	X₁	X₂	X₃
Y	1.000	0.357	0.620	0.518
X₁	0.357	1.000	0.321	0.477
X₂	0.620	0.321	1.000	0.539
X₃	0.518	0.477	0.539	1.000

TABLE 5.1 Intercorrelation Matrix.

The equation employed in the Kelley-Salisbury technique is this:

$$\begin{array}{ccc} R & \beta_z & V \end{array}$$

$$\begin{bmatrix} 1.000 & 0.321 & 0.477 \\ 0.321 & 1.000 & 0.539 \\ 0.477 & 0.539 & 1.000 \end{bmatrix} x \begin{bmatrix} \beta_{z_1} \\ \beta_{z_2} \\ \beta_{z_3} \end{bmatrix} = \begin{bmatrix} 0.357 \\ 0.620 \\ 0.518 \end{bmatrix}.$$

---

[10] Walker, H. M. & Lev, J. (1953). *Statistical Inference*. New York: Rinehart and Winston, Inc.

In general, the equation $\beta_z = R^{-1}V$ has a solution only if all the predictors are independent of one another. The iterative technique that is the heart of the Kelley-Salisbury method, however, will find a solution even when the predictors are statistically dependent.

We begin by choosing arbitrary values for the vector β_z of beta weights. We then perform one matrix multiplication, $R\,\beta_z = V_r$, where V_r is a vector of trial validities. The difference of largest magnitude between elements of V and elements of V_r is then added to the corresponding trial beta weight in the vector β_z, and then this process is repeated over again.

In our example, we set the weights equal to the validities. Our first multiplication is

$$R\,\beta_z = \begin{bmatrix} 1.000 & 0.321 & 0.477 \\ 0.321 & 1.000 & 0.539 \\ 0.477 & 0.539 & 1.000 \end{bmatrix} x \begin{bmatrix} 0.357 \\ 0.620 \\ 0.518 \end{bmatrix} = \begin{bmatrix} 0.803 \\ 1.014 \\ 1.022 \end{bmatrix} = V_r,$$

then

$$V - V_r = \begin{bmatrix} 0.357 \\ 0.620 \\ 0.518 \end{bmatrix} - \begin{bmatrix} 0.803 \\ 1.014 \\ 1.022 \end{bmatrix} = \begin{bmatrix} -0.446 \\ -0.394 \\ -0.504 \end{bmatrix}.$$

The difference of largest magnitude is $V_3 - V_{r_3} = -0.504$, so this amount is added to β_{z_3}.

The vector of approximate weights is now

$$\beta_z = \begin{bmatrix} 0.357 \\ 0.620 \\ 0.014 \end{bmatrix}.$$

This process is repeated, adjusting one element of β_{z_3} at each iteration until V_r converges on V. In our example, this conversion occurs on the 18th iteration, as detailed in Table 5.2.

Approximations to β_z			Vector of Trial Validities		
β_1	β_2	β_3	V_1 **0.357**	V_2 **0.620**	V_3 **0.518**
0.357	0.620	0.518	0.803	1.014	1.022
0.357	0.620	0.014	0.563	0.742	0.518
0.151	0.620	0.014	0.357	0.676	0.420
0.151	0.620	0.112	0.403	0.729	0.518
0.151	0.511	0.112	0.368	0.620	0.459
0.151	0.511	0.171	0.397	0.652	0.518
0.111	0.511	0.171	0.357	0.639	0.499
0.111	0.492	0.171	0.350	0.620	0.489
0.111	0.492	0.200	0.364	0.635	0.518
0.111	0.477	0.200	0.360	0.620	0.510
0.111	0.477	0.208	0.363	0.625	0.518
0.105	0.477	0.208	0.357	0.623	0.515
0.105	0.474	0.208	0.356	0.620	0.514
0.105	0.474	0.212	0.358	0.622	0.518
0.105	0.474	0.212	0.358	0.620	0.516
0.105	0.472	0.214	0.359	0.621	0.518
0.103	0.422	0.214	0.3566	0.6204	0.5175
0.103	0.472	0.215	0.3571	0.6209	0.5185
0.103	0.471	0.215	0.3567	0.6199	0.5180

[a]The final values of the weights are $\beta_{z1} = 0.103$, $\beta_{z2} = 0.471$, and $\beta_{z3} = 0.215$.

TABLE 5.2 The Iterations of the Kelley-Salisbury Method.

5.2.2 Grouping to Optimize Judges' R^2

Each judge, in JAN, is associated with their individual multiple linear regression equation. R^2, for this equation, is the judge's predictive efficiency in making their final judgments for each pattern in the profile. If a judge has been predicted to be "extremely low" for their R^2 value, then that judge should be dropped from the study.

Consider starting JAN with the assumption that all the judges are using unique judgmental policies. Then the judges are acting independently of one another in making their judgments. Assume that we have three judges, each judging a profile matrix with four independent variables. The profile matrix, X, of $n = 5$ observations or cases is paired one-to-one with the judges ratings or rank judgment vector, $\vec{Y}$. In this case, the JAN can be thought of as being based upon the following full regression model run:

$$\text{For each judge, we have: } \vec{Y_i} = \begin{bmatrix} y_{i1} \\ y_{i2} \\ y_{i3} \\ y_{i4} \\ y_{i5} \end{bmatrix} \text{ and } X = \begin{vmatrix} x_{1,1} & x_{1,2} & x_{1,3} & x_{1,4} \\ x_{2,1} & x_{2,2} & x_{2,3} & x_{2,4} \\ x_{3,1} & x_{3,2} & x_{3,3} & x_{3,4} \\ x_{4,1} & x_{4,2} & x_{4,3} & x_{4,4} \\ x_{5,1} & x_{5,2} & x_{5,3} & x_{5,4} \end{vmatrix}.$$

where y_{ij} is the ith judge's rating or ranking for the jth independent variable, and $x_{i,j}$ is the jth independent variable's value viewed by the jth judge.

Using each judge's model, the regression full model, for all judges as a singleton cluster would be:

$$\begin{vmatrix} Y_1 \\ Y_2 \\ Y_3 \end{vmatrix} \text{ with the new profile matrix } \begin{vmatrix} X & \dot{0} & \dot{0} \\ \dot{0} & X & \dot{0} \\ \dot{0} & \dot{0} & X \end{vmatrix} \text{ where } \dot{0} = \begin{vmatrix} 0000 \\ 0000 \\ 0000 \\ 0000 \\ 0000 \end{vmatrix}.$$

The R^2 for this new matrix and vector is the predictive efficiency for the clustering of all singleton clusters. On all iterations of the JAN, the R^2 for the new pseudo-judge is determined by replacing the two judges or pseudo-judges which requires building a new judgmental vector and profile matrix. This construction is illustrated for placing J_1 and J_2 in the same cluster: the vector $\begin{vmatrix} Y_{1\&2} \\ Y_2 \end{vmatrix}$, where $Y_{1\&2}$ *is the column vector with* 10 *rows* – $Y_1's$ *followed by* $Y_2's$, and $\begin{vmatrix} X_{1\&2} & \dot{0}_{10,4} \\ \dot{0}_{5,4} & X \end{vmatrix}$. Note that $X_{1\&2}$ is the 10 by 4 matrix containing one copy of X in the first five rows followed by X in the second row. $\dot{0}_{i,j}$ is a zero matrix with i rows and j columns.

5.2.3 Alternative Method for JAN

A related methodology for the JAN study in this chapter could be completed as outlined by Harvill, Lang, & McCord.[11] This methodology is called

---

[11] Harvill, L. M., Lang, F., & McCord, R. S. (2004). *Determining Component Weights in a Communications Assessment Using Judgmental Policy Capturing*. Medical Education Online [serial outline], 9:12 available from http://www.med-ed-online.org.

judgmental policy capturing (JPC) which is used to determine component weights for complex assessment components in medical education:

JPC is used to capture policy from individual judges relative to various profiles of assessment outcomes and, when averaged, can provide weights for the components of the assessment. It is not used to look at an individual case or "test item" and is not used to determine a minimum passing score or set a standard for a particular assessment instrument as is the case with standard setting methods.

The specific steps in the procedure for a single iteration of JPC are:

1. Judges are asked to independently provide overall ratings of the performance of examinees on a complex skill assessment often using graphic representations or profiles of the scores from those assessments for a large number of examinees. The judges' ratings are often couched in terms of the competence of the examinees. Repeated ratings of some of the same profiles provide a measure of stability of the judgments of each individual judge although this is not a mandatory art of the process. Overall ratings can also be compared among the various judges to determine the degree of agreement among the judges but consensus among the judges is not a goal of this process.

2. Multiple regression analysis is used to determine appropriate regression or beta weights for the assessment components using the various assessment component scores as the predictor or independent variables and the judge's overall ratings for the assessment profiles as the dependent variable or variable to be predicted for each judge. This is a widely used statistical procedure that provides a means for capturing how each expert valued each skill component in arriving at his/her global ratings of performance.

3. The regression weights from each of the judges can be expressed as percentages and, thus, provide a straightforward statement about the relative importance of each assessment component in determining the overall ratings made by that judge. If the assessment has three facets, one judge's percentage weights for those components might be 30 percent, 35 percent, and 35 percent, respectively, while a second judge's weights or values for the importance of the three facets in terms of overall performance might be 40 percent, 40 percent, and 20 percent, respectively. A third expert's percentage weights might be 35 percent, 40 percent, and 25 percent, respectively.

4. The sets of weights provided by the judges are then typically averaged to arrive at a composite set of weights for that particular skill assessment. For the example above, the average weight for the facets would be 35 percent, 38.33 percent, and 26.67 percent, respectively. This set of weights could then be applied to new assessments of that skill to provide an overall rating or score for examinee performance. For example, if an examinee received percentage scores of 65 percent, 85 percent, and 80 percent on the three assessment components, the examinee's overall percentage score would be:

$$(.35 \times 65) + (.3833 \times 85) + (.2667 \times 80) = 76.67 \text{ percent.}$$

5. The process can be continued with an additional iteration of the above steps after the panel of judges has had some opportunity to see their own set of regression weights and those of the other panel members; some discussion among the panel members might or might not be included.

5.3 JUDGMENTAL ANALYSIS IN RESEARCH

The field of education contains many examples of JAN. The extent to which graduate school faculty held a single admissions policy was examined by Houston[12] and also by Williams, Gab, and Lindem[13]. Roscoe and Houston[14] looked at the relevance of the Graduate Record Examination (GRE) as an admission standard for doctoral study at Colorado State College. There were a number of studies that dealt with teacher effectiveness. Houston and Bentzen[15] looked at teaching effectiveness in culturally deprived junior high math classes. Houston and Boulding[16] used JAN to capture the teacher effectiveness policies of College of Education faculty.

---

[12] Houston, S. R. (1968). Generating a projected criterion of graduate school success via normative analysis. *Journal of Experimental,* 37, 53-58.

[13] Williams, J. D., Gab, D. & Linden, A. (1969). Judgmental analysis for assessing doctoral admissions programs. *Journal of Experimental Education*, 38(2), 92-96.

[14] Roscoe, J. T. & Houston, S. R. (1969). The predictive validity of GRE scores for a doctoral program in education. *Educational and Psychological Measurement, 29,* 507-509.

[15] Houston, S. R. & Bentzen, M. M. (1969). Teaching effectiveness in culturally deprived junior high mathematics classes. *Journal of Experimental Education,* 38(1), 73-78.

[16] Houston, S. R. & Boulding, J. T. (1974). Capturing faculty teaching effectiveness policies with judgmental analysis. *California Journal of Educational Research, 25,* 134-139.

In a study by Houston, Crosswhite, and King,[17] a modified form of JAN, called JAN-B, was used as a means for capturing a group or collective teacher effectiveness policy from selected students. In JAN-B, a common group is joined together to form one judge. For example, in their study, students were asked to rank teachers on nine variables. The students were then grouped together and the groups were used as individual judges in the JAN analysis. One of the groupings consisted of five judges: freshman, sophomore, junior, senior, and graduate students. JAN was also used to identify a policy of rated school effectiveness in the experimental League of Cooperating Schools project (Houston, Duff, and Roy.)[18] Leonard, Gruetzemacher, Wegner, and Whittington[19] used JAN to evaluate the college of education and psychology at a state supported university. The evaluation policies of citizens and parents, clergy, lay educators, and religious educators were the focus of a study by Leonard, Gruetzemacher, Maddox, and Stewart.[20] Other studies in education that utilized the JAN technique included the development of a learning disability construct (Mabee),[21] an educational decision-making situation in special education (Stock),[22] and faculty policies for granting salary increases (Williams, Mabee, and Brekke).[23] Education was not the only discipline to use the JAN method. Holmes and Zedeck[24] used JAN to capture the policies involved in assessing paintings with respect to beauty. There were also studies that used JAN to identify

---

[17] Houston, S. R., Crosswhite, C. E., & King, R. S. (1974). The use of judgment analysis in capturing student policies of rated teacher effectiveness. *Journal of Experimental Education, 43*(2), 28-34.

[18] Houston, S. R., Duff, W. L. Jr., & Roy, M. R. (1972). Judgment analysis as a technique for evaluating school effectiveness. *Journal of Experimental Education, 40*(4), 56-61.

[19] Leonard, R. L., Gruetzemacher, R. R., Wegner, W., & Whittington, B. (1980). Judgment analysis for evaluating a college. *Journal of Experimental Education, 49*(1), 38-44.

[20] Leonard, R. L., Gruetzemacher, R. R., Maddox, V. A., & Stewart, D. K. (1982). Evaluation policy definition by judgment analysis among archdiocesan school constituents. *Journal of Experimental Education, 50*(4), 205-210.

[21] Mabee, W. S. (1978). An investigation of the learning disability construct by the JAN technique. *Journal of Experimental Education, 46*(4), 19-24.

[22] Stock, G. C. (1969). *Judgmental analysis for the educational researcher.* Unpublished doctoral dissertation, University of Northern Colorado, Greeley.

[23] Williams, J. D., Mabee, W. S., & Brekke, K. A. (1976). Faculty policies for granting salary increases. *Journal of Experimental Education, 45*(2), 65-69.

[24] Holmes, G. P., & Zedeck, S. (1973). Judgment analysis for assessing paintings. *Journal of Experimental Education, 41*(4), 26-30.

pornographic material (Houston and Houston;[25] Houston, Houston, and Ohlson).[26] Zedeck and Kafry[27] used JAN to cluster raters who had similar strategies in a study involving nurses.

5.4 EXAMPLE JAN STUDY

Consider the following study, which was generated by a research request from the College of Business Administration at Omega University. The Information Systems Department wanted to find and measure a set of criteria that would be descriptive of the doctoral candidate selection process by the IS faculty. This set of criteria would be descriptive of the candidates' success in the program. Selective procedures for doctoral candidates in IS involve looking at many predictor variables; some of these variables are GRE population, GMAT scores, Miller Analogies Test (MAT) scores, Strong Vocational Interest Blank (SVIB), incoming GPA, and previous major field of study. The MMPI is also required for each incoming doctoral candidate. However, in the past, many of the doctoral graduates did not take the MMPI. Because of this, the 13 subscales of the MMPI were not included in the study. To describe the doctoral candidate selection process used by the IS faculty, three questions had to be answered:

- How many distinct judgmental policies are present among the IS faculty with respect to what they consider the important cognitive and affective skills of an outstanding graduate student?

- Which of the affective and cognitive skills being considered for each graduate do the faculty deem most important?

- Which predictor variables used in determining admission to the program relate to the captured judgmental policy or policies?

5.4.1 Statement of Problem

At present there seems to be a wide variety of methodologies available to capture and analyze judgmental policies where either the investigator is

[25] Houston, J. A., & Houston, S. R. (1974). Identifying pornographic materials with judgment analysis. *Journal of Experimental Education, 42*(4), 18-26.

[26] Houston, J. A., Houston, S. R., & Ohlson, E. L. (1974). On determining pornographic material. *The Journal of Psychology, 88*, 277-287.

[27] Zedeck, S., & Kafry, D. (1977). Capturing rater policies for processing evaluation data. *Organizational Behavior and Human Performance, 18*, 269-294.

looking only at predictor variables alone or there is one criterion variable and several predictor variables. Missing in the literature is a methodology for capturing and analyzing judgmental policies where the investigator is looking at a set of criterion variables and a set of predictor variables simultaneously. Therefore, the problem is to attempt to capture and analyze judgmental policies of the IS faculty while looking at both a set of criterion variables and a set of predictor variables.

5.4.1.1 Purpose

The basic purpose of this study, which is service oriented in nature, is to develop a methodology for capturing and analyzing judgmental policies where the investigator is looking at a set of criterion variables and a set of predictors in conjunction with each other. This type of methodology is needed so that the technique can be used to a great degree in judgmental research.

The questions answered can be subdivided into two general populations. The first population of concern pertains to how well the methodology answered the questions of the IS research request. These questions are as follows:

- How many different judgmental policies exist within the IS Department concerning what constitutes an outstanding IS doctoral graduate?

- Which of the affective and cognitive skills being considered for each graduate do the faculty deem most important?

- What kinds of information that they now collect are important indicators of a graduate's ability to succeed as an IS employee?

The second population of concern involves interpretation of computer outputs. The major questions that can be considered are as follows:

- What types of judgmental policies are captured?

- Is it possible to determine whether or not each judge is a "good" judge?

- How many judgmental policies are captured?

- What are the most important predictor and criterion variables?

- What is the relationship between the criteria and predictors of importance?

- Is the procedure for analyzing a captured policy an intuitive or a statistical process?

**The Criterion Instrument Used to Obtain the Ratings
of the Faculty forthe 50 Doctoral Graduates
with Whom They Were Familiar**

Project—Can future success of doctoral candidates be predicted with a high degree of certainty using only a few predictors?

Directions

      On an attached sheet you will find a list of the doctoral graduates from the IS Department. As judges (i.e., faculty of the IS Department), please rate the graduates whom you feel qualified to rate. You will be asked to rank them in nine different categories, plus a general overall rank. When evaluating the graduates, please rate them in comparison to the other graduates with whom you are familiar and in terms of what you think constitutes an outstanding IS graduate.

Please rank the graduates from 1 to 5, where 1 = poor, 2 = below average, 3 = average, 4 = above average, and 5 = outstanding, on the following categories.

Name of Graduate_____ Name of Judge_____

_____

Section 1 (pertains to the graduate while a student in the IS Department).

RANK

_____1. Academic success (final GPA, comps, orals, and dissertation)

_____2. Verbal ability or verbal articulation

_____3. Intellectual ability

_____4. Interpersonal relationships—faculty and supervisors

_____5. Interpersonal relationships—peers

_____6. Interpersonal relationships—to people he is serving (students, inmates, etc.)

_____7. Leadership initiative in getting things done (projects, meetings, etc.)

_____8. Personal characteristics (emotionally mature, friendly, enthusiastic)

_____9. Improvement of self (personally and professionally)

_____10. General overall rank combining all categories above into a single rank from 1 to 5

FIGURE 5.1 Study Criterion Instrument.

There are a few limitations to this study:

- The subject to predictor ratio should have been larger.
- This specific study cannot be generalized beyond the IS Department at Omega University.

5.4.2 Predictor Variables

The following hierarchy of predictor variables was used in the study. This information was obtained either from the faculty of the IS Department or the Omega University Graduate Office. As mentioned earlier, this information was used in the selection of IS doctoral students.

The cognitive variables are the following:

X_2 = GRE population score in social sciences

X_3 = GRE population score in the humanities

X_4 = GRE population score in the natural sciences

X_5 = GMAT verbal score

X_6 = GMAT qualitative score

X_7 = GMAT composite score

X_8 = Incoming GPA earned in student's master's degree program (A = 4.0, B = 3.0, C = 2.0, D = 1.0)

X_9 = Miller Analogies Test score

Note: Variables X_2, X_3,..., X_9 are in raw score form.

The affective variables, from Group 5 of the SVIB, are the following:

X_{10} = Interest score for becoming an IT manager

X_{11} = Interest score for becoming a database manager

X_{12} = Interest score for becoming a systems analyst

X_{13} = Interest score for becoming a programmer

X_{14} = Interest score for becoming a Website designer

X_{15} = Interest score for becoming a CIO

The biographic variables are the following:

X_{16} = Age when admitted to the IS doctoral program

X_{17} = 1 if the master's degree program was in IS or MIS; 0 otherwise

There are eight scores on the previous version of the SVIB Group 5 section, occupational interest. This inventory has been revised three times in the last 10 years and new occupational classifications have been added. Because of this, only those occupational populations of interest that are on all editions were used in the study. Also, the scores on the same year edition were not rescaled. This decision was made after consulting with the director of research marketing.

The means, standard deviations, and raw scores for the predictor variables X_2, X_3,..., X_{17} are considered confidential information by the faculty of the IS Department. Because of this, they were not included in the study.

5.4.3 Criterion Variables

The following hierarchy of criterion variables was used in the study.

The cognitive variables are the following:

Y_1 = Academic success (final GPA, comprehensives, orals, and dissertation)

Y_2 = Verbal ability or verbal articulation

Y_3 = Intellectual ability

The social or affective variables are the following:

Y_4 = Interpersonal relationships with faculty and supervisors

Y_5 = Interpersonal relationships with his peers

Y_6 = Interpersonal relationship with the people he is serving (students, inmates, patients, etc.)

Y_7 = Leadership initiative in getting things done (projects, etc.)

Y_8 = Personal characteristics (emotionally mature, friendly, enthusiastic)

Y_9 = Improvement of self (personally and professionally)

The composite variable is the following:

Y_{10} = General overall rank combining all of the above nine categories into a single rank

Each of the criteria was scored on a scale from 1 to 5 where 1 = poor, 2 = below average, 3 = average, 4 = above average, and 5 = outstanding.

The criterion instrument included in the faculty questionnaire was developed by three members of the IS faculty. Considerations for the instrument were based on the following question: "If it were my job to hire a PhD with a degree in IS work, what characteristics or attributes would I would expect him to have?" From this question, the nine criterion variables were generated and used in the study. Most of the IS doctorates have graduated in the past three years and so consideration was not given to the number of publications, employment effectiveness in their present job, and professional affiliations.

The criterion instrument was submitted to the 16 members of the IS faculty with the following instructions: Please rate each of the above PhD graduates with whom you are familiar on a scale from 1 to 5 where 1 = poor, 2 = below average, 3 = average, 4 = above average, and 5 = outstanding. The basis for comparison should be your conception of what constitutes an outstanding PhD graduate in IS.

Notice that this is an ipsative type of JAN ranking. Approximately 500 ratings were obtained through the distribution of the questionnaire.

5.4.4 Questions Asked

- How many of the judges adhered strictly to the profile scores when ranking the IS doctoral graduates?

- How many different judgmental policies existed?

- Which predictor variables, $X_2, X_3, ..., X_{17}$, were the most effective in redirecting the obtained judgmental policy?

5.4.5 Method Used for Organizing Data

The following output data for the normative JAN procedure was presented and summarized in Tables 5.3 to 5.7:

- The means and standard deviations for the profile scores and the judgments

- The correlation matrix for the profile scores

- The correlation matrix for the judgments

- The correlations between the judgments and the profile scores

- The sequential and cumulative R^2 drops, the judges combined, and the single member systems remaining in each step of the JAN procedure

The means and standard deviations for the profile scores and judgments are listed in Table 5.3. The mean for each of the profile scores was approximately 3. Because each Y_i value, where $i = 1,..., 9$, only takes on values from 1 to 5, and because the 40 sets of ratings were selected at random, these

Profile	Mean	Standard Deviation
Y_1 = Academic success (final GPA, comps, orals,dissertation)	3.27	1.18
Y_2 = Verbal ability or verbal articulation	3.15	1.37
Y_3 = Intellectual ability	3.20	1.17
Y_4 = Interpersonal relationships—faculty and supervisor	3.10	1.14
Y_5 = Interpersonal relationships—peers	3.07	1.06
Y_6 = Interpersonal relationships—to people he is serving (students, inmates, etc.)	2.92	1.10
Y_7 = Leadership initiative in getting things done (projects, meetings, etc.)	3.02	1.23
Y_8 = Personal characteristics (emotionally mature, friendly, enthusiastic)	3.27	1.22
Y_9 = Improvement of self (personally and professionally)	3.15	1.31
Judgments		
Judge 1	3.15	1.64
Judge 2	3.10	2.05
Judge 3	3.07	1.69
Judge 4	3.60	1.50
Judge 5	3.90	1.91
Judge 6	3.57	1.41
Judge 7	3.40	1.56
Judge 8	3.97	1.80
Judge 9	3.40	1.60
Judge 10	3.45	1.82
Judge 11	3.82	1.80
Judge 12	3.47	1.39

TABLE 5.3 Means and Standard Deviations for the Profile Scores and the Judgments.

results seem consistent. The standard deviations also seem consistent with what should be expected. (Note that 12 judges returned the instrument, as discussed later.)

The means for the judgments all lie between 3 and 4. This indicates that the judges tended to put more of the 40 sets of ratings in Groups 1 through 3 rather than in Groups 5 through 7. The standard deviations of the judgments were also consistent with what should be expected.

The correlation matrix for the profile scores, the correlation matrix for the judgments, the intercorrelation matrix between the profile scores and the judgments, and the initial R^2 for each of the judgments are included in Tables 5.4 through 5.7, respectively.

Variable	1	2	3	4	5	6	7	8	9
1									
2	0.79								
3	0.85	0.84							
4	0.01	0.07	0.02	0.01					
5	0.06	0.08	0.03	0.84					
6	0.07	0.37	0.21	0.76	0.75				
7	0.42	0.72	0.53	0.39	0.42	0.62			
8	0.21	0.35	0.22	0.72	0.70	0.66	0.54		
9	0.33	0.54	0.42	0.50	0.51	0.59	0.77	0.75	

TABLE 5.4 Correlation Matrix for the Profile Scores.

The correlation matrix for the profile scores, shown in Table 5.4, indicates that the cognitive variables Y_1, Y_2, and Y_3 all correlate highly with one another. Variables Y_4, Y_5, and Y_6 correlate highly with one another but quite low with variables Y_1, Y_2, and Y_3. Because variables Y_4, Y_5, and Y_6 are affective variables, this was to be expected. Variables Y_7, Y_8, and Y_9 are also affective variables, but their interpretation is not quite as clear. Y_8 appears to correlate the highest with Y_4, Y_5, Y_6, and Y_9. Y_7 and Y_9 correlate highly with each other but also fairly high with variables Y_1 through Y_6 and Y_8. It appears that Y_1 through Y_3, Y_4 through Y_6, and Y_8 through Y_9 should be grouped into three separate clusters. A factor analysis was completed on the profile scores and confirmed the three clusters.

Variable	1	2	3	4	5	6	7	8	9	10	11	12
1												
2	0.90											
3	0.92	0.83										
4	0.82	0.37	0.73									
5	0.92	0.92	0.90	0.79								
6	0.87	0.90	0.83	0.73	0.89							
7	0.84	0.86	0.86	0.82	0.88	0.85						
8	0.85	0.85	0.84	0.71	0.85	0.81	0.86					
9	0.84	0.91	0.82	0.72	0.90	0.89	0.87	0.79				
10	0.86	0.89	0.85	.80	.87	.80	0.87	.85	0.81			
11	0.93	0.91	.89	0.88	0.92	0.85	.91	0.87	0.86	0.92		
12	0.92	0.90	0.88	0.85	0.90	0.85	0.88	0.88	0.85	0.91	0.95	

TABLE 5.5 Correlation Matrix for the Judgments.

The correlation matrix for the judgments, shown in Table 5.5, indicates a high agreement among the judges in the ratings that each gave to the 40 sets of ratings. This will also be indicated later in the analysis of the JAN procedure.

The intercorrelation matrix between the profiles and the judgments, shown in Table 5.6, indicates that Judges 4, 7, and 10 have somewhat higher relative correlations with variables Y_1, Y_2, and Y_3 than do the rest of the judges when the correlations are examined for each judge over all the profiles. Again, this result shows up in the analysis of the JAN procedure.

On the criterion instrument, there were 480 usable sets of ratings obtained that could be used in the study. One of the judges was eliminated from the study. From these 480 sets of ratings, 40 sets of ratings were randomly selected. The modified Q-sort instrument with the necessary instructions was then distributed to the 13 faculty members of the IS faculty and to the 3 past members of the IS faculty. Instruments were returned by 12 of the 16 faculty members.

	Judgments											
Variables	1	2	3	4	5	6	7	8	9	10	11	12
1	0.45	0.39	0.97	0.62	0.35	0.31	0.47	0.39	0.33	0.37	0.57	0.54
2	0.58	0.56	0.56	0.71	0.56	0.46	0.65	0.61	0.45	0.74	0.74	0.74
3	0.50	0.40	0.40	0.73	0.43	0.32	0.54	0.46	0.39	0.64	0.62	0.61
4	0.75	0.71	0.71	0.47	0.73	0.81	0.61	0.68	0.75	0.59	0.64	0.64
5	0.69	0.78	0.73	0.43	0.77	0.78	0.64	0.67	0.79	0.59	0.61	0.62
6	0.82	0.85	0.80	0.57	0.83	0.84	0.77	0.81	0.79	0.75	0.75	0.75
7	0.76	0.76	0.74	0.75	0.78	0.67	0.78	0.68	0.68	0.81	0.85	0.82
8	0.72	0.76	0.79	0.60	0.77	0.78	0.75	0.73	0.78	0.66	0.77	0.80
9	0.76	0.75	0.76	0.68	0.79	0.68	0.70	0.65	0.72	0.73	0.82	0.85

Profiles (row label, left side)

TABLE 5.6 Intercorrelation Matrix Between the Profiles and the Judgments.

The initial R^2 for each of the judges ranged from 0.9735 to 0.8134. This indicates that each judge paid strict attention to the variables upon which the ratings were based. Because of the excellent job done by each of the judges on the modified Q-sort, they were all included in the study. The sequential and cumulative R^2 drops, the judges combined, and the single-member systems remaining in each step of the Type A normative JAN procedure are listed on Table 5.7.

Judges	RSQ
Judge 1	0.9282
Judge 2	0.9521
Judge 3	0.8708
Judge 4	0.8134
Judge 5	0.9212
Judge 6	0.8860
Judge 7	0.8761
Judge 8	0.8436
Judge 9	0.8896
Judge 10	0.9109
Judge 11	0.9719
Judge 12	0.9735

TABLE 5.7 Initial R^2 for Each of the Judges.

5.4.6 Subjects Judged

The IS Department had its first doctoral graduate 10 years earlier. Since then, not including the previous year's graduates, 55 individuals have received their doctoral degree from the IS Department. Of these 55 graduates, it was possible to get complete information on the predictor variables for 50 of them. These 50 were the individuals used in the study.

5.4.7 Judges

At present, there are 13 faculty members of the IS Department who have worked with some or all of the 50 doctoral graduates. There were three faculty members on the staff of the IS Department over the 10 years who have left for positions in other universities. Because they are familiar with many of the graduates being studied, they were included in the study. These 16 past or present faculty members were used as judges.

5.4.8 Strategy Used for Obtaining Data

Procedure 1:

Out of the approximately 500 sets of ratings obtained on the criterion instrument, 40 were randomly selected.

Procedure 2:

Each of the 16 members of the IS faculty were asked to do a modified Q-sort on the 40 sets of ratings. Each set of ratings was called a profile. They were asked to compare the 40 profiles against one another and place them in seven groups. Group 7 would represent those profiles they considered the best of the group of 40 and Group 1 would represent those profiles they considered the poorest in the group of 40. Groups 2 through 6 lay on an ordinal scale between one and seven. The only restrictions were that each judge must place at least one profile in each of the seven groups. Hopefully, this forced the 12 judges to think along the whole continuum. The instrument and the instructions used for the Q-sort were presented to the IS faculty in a training session before they actually conducted this part of the study.

Procedure 3:

After the judges completed the rankings, a Type A JAN was run to determine the number of judgmental policies. The decision as to the number of policies existing was based on the following consideration: A judge who was unable to identify at least one significant factor is failing to relate any predictor variable set to any criterion variable.

Procedure 4:

One judgmental policy was determined and a regression model was obtained where the rankings of the 40 profiles scored by each of the judges was the criterion Y_{11}, and $Y_1, ..., Y_9$ were the predictors. The systematic procedure was used to determine which of the nine criteria the IS faculty considered most important in an outstanding doctoral graduate. This process is visually displayed in the regression flowchart shown in Figure 5.2. If the subset of variables removed in the restricted model does not constitute an R^2 drop from the full model of 0.05 or more, then this subset of variables was not considered to make a significant contribution.

Procedure 5:

The average score for each IS doctoral graduate was determined for each of the criteria, Y_i, $i = 1, 2,..., 9$. This score on each criterion was determined by averaging all the scores the graduate received from those judges who rated him. Average scores on each of the criteria along with the predetermined weight for the criteria were used to obtain a predicted score, Y_{12}, for each of the IS graduates: $Y_{12} = b_1 Y_1 + b_2 Y_2 + ... + b_9 Y_9$.

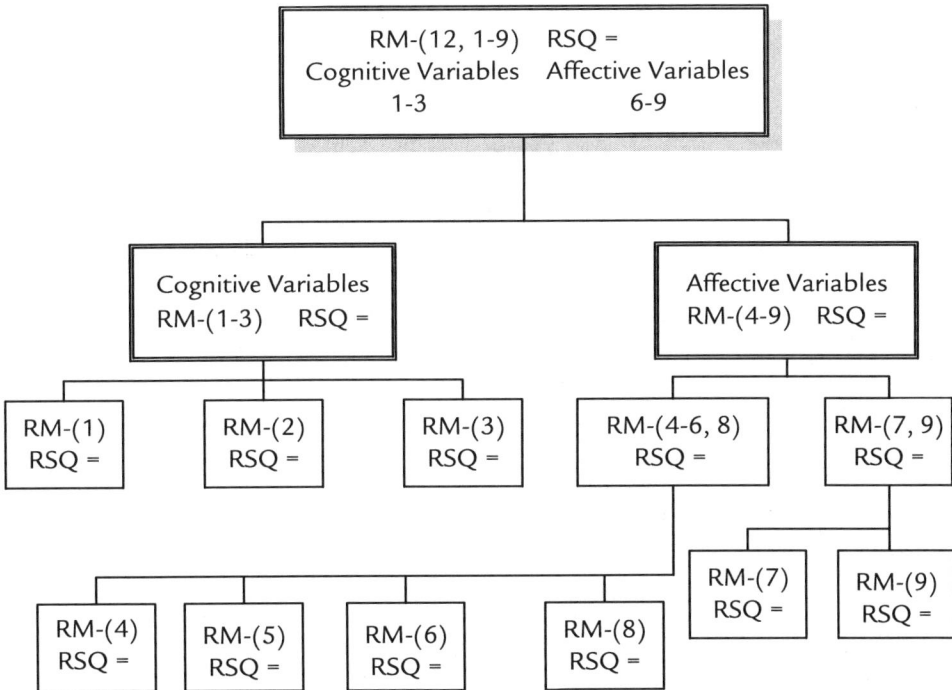

Note: Variables 1, 2, 3, … , 9, 12
       12, 1-9 means a model using 12 as the criterion and 1-9 as predictors
       RM-( ) represents the restricted model formed by dropping variables ( ) out
       of the full model

FIGURE 5.2 Regression Chart for the Criterion Variables.

The second main objective of this methodology was to determine which of the predictors used for admittance to the IS doctoral program are of the greatest importance. Because a judgmental policy was already determined in regard to what constitutes an outstanding IS graduate, this was now possible.

Procedure 6:

A regression model was run using the predicted score from Procedure 5 as the criterion, Y_{12}, and X_2, X_3,..., X_{17} were used as predictors. The systematic procedure that was used to determine which of the predictors used for admittance to the IS doctoral program are of greatest importance is given in the regression flowchart in Figure 5.3. If the subset of variables removed in the restricted model does not constitute an R^2 drop from the full model of 0.05 or more, then this subset of variables was not considered to make a significant contribution.

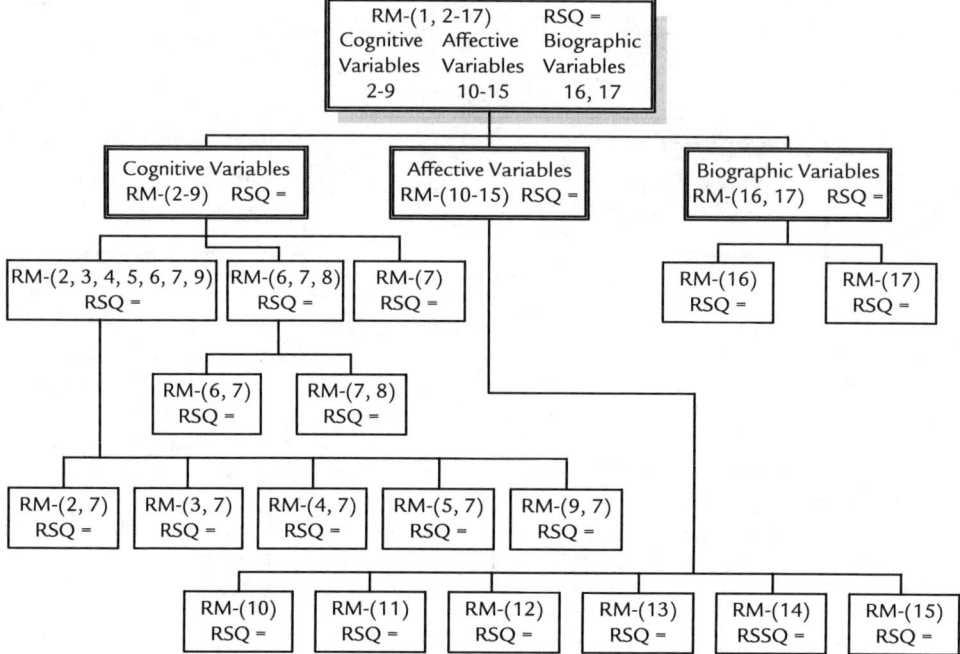

FIGURE 5.3 Regression Chart for the Predictor Variables.

5.4.9 Checking the Model

Checking the model is verifying that the variables of the final model do not violate assumptions upon which the model is based. This involves careful determination that the variables are correctly distributed and standard errors are random, as well as looking at any other stringent variate requirements.

The purpose of checking the model is to make sure that the assumptions underlying the statistical tests used have not been violated. In some experimental designs, we may reduce the variance of the estimator and increase the information by either reducing the variance (noise) or increasing the sample size (volume of the signal).

In practice, the assumptions for a linear model are rarely satisfied. The experimenter does not know all of the important variables in a process nor does he know the true functional relationships. Therefore, the function chosen to fit the true relation is only an approximation, and the variables included in the experiment form only a subset of the total. The random error is thus a composite of error caused by the failure to include all of the important process variables as well as the error in approximating the function.

To determine the number of judgmental policies present, the factors previously mentioned were used. Because the original R^2 drop is only 0.0729, this indicates that one judgmental policy provided does not exhibit a large R^2 drop from one grouping to the next. Further examination of the R^2 drop from one grouping to the next, as shown in Table 5.8, indicates a linear trend from one stage to the next. On the basis of this evidence, it was decided that one judgmental policy existed for the 12 judges. Because of

Methodology for Criterion = 11				
Stages	**Judges**	**R^2**	**Drop in R^2**	**Collective Drop in R^2**
1	1, 2, 3, 4, 5, 6, 7, 8, 9, 10, 11, 12	0.9084		
2	(3, 9), 1, 2, 4, 6, 7, 8, 9, 10, 11, 12	0.9065	0.0019	
3	(3, 5), (6, 9), 1, 2, 4, 7, 8, 10, 11, 12	0.9041	0.0024	0.0043
4	(3, 5), (6, 9), (11, 12), 1, 2, 4, 7, 8, 10	0.9013	0.0028	0.0071
5	(3, 5), (6, 9), (7, 10), (11, 12), 1, 2, 4, 8	0.8974	0.0038	0.0109
6	(1, 2), (3, 5), (6, 9), (7, 10), (11, 12), 4, 8	0.8930	0.0044	0.0154
7	(1, 2), (3, 5, 8), (6, 9), (7, 10), (11, 12), 4	0.8884	0.0046	0.0199
8	(1, 2), (3, 5, 8), (6, 9), (4, 7, 10), (11, 12)	0.8831	0.0053	0.0253
9	(1, 2), (3, 5, 8, 6, 9), (4, 7, 10), (11, 12)	0.8771	0.0060	0.0313
10	(1, 2), (3, 5, 8, 6, 9, 11, 12), (4, 7, 10)	0.8684	0.0087	0.0400
11	(1, 2, 3, 5, 8, 6, 9, 11, 12), (4, 7, 10)	0.8565	0.0119	0.0519
12	(1, 2, 3, 5, 8, 6, 9, 11, 12, 4, 7, 10)	0.8533	0.0210	0.0729

TABLE 5.8 Stages of the JAN Procedure for the Judges.

this, it was not necessary for the IS faculty to meet and arbitrate a single policy.

A point of interest that was mentioned previously is again apparent when one examines stages 11 and 12. Judges 4, 7, and 10 are grouped together in stage 11 and when combined with the other nine judges in stage 12, we have the largest R^2 drop. Although this drop was not large enough to warrant two judgmental policies, it was consistent with the results found earlier in the examination of previous tables.

A regression model Y_{11} on Y_1, Y_2,..., Y_9 was completed to determine what the nine criteria judges deemed most important if they were to hire a PhD with a degree in IS. Here we were trying to determine what the faculty considered the highest priority of the nine criteria they considered important for an outstanding doctoral graduate to possess. The schematic for the flowchart was developed by an examination of the correlations between the criteria, Y_1, Y_2,..., Y_9, and factor analysis.

5.4.10 Extract the Equation

Extracting the equation derives the formula that explains the change in the dependent variable. This formula can be used to predict future values of the dependent variable or to analyze the sources of variation.

In a predictive regression study, the last step is to extract the equation for the purpose of predicting future and current values of the dependent variable.

Type A normative JAN analysis was completed to determine the number of judgmental policies that existed among the IS faculty. Only one judgmental policy existed and arbitration among the judges was not necessary. After this, two regression analyses were made to determine the important criterion and/or predictors as related to the judgmental policy of the IS faculty. A complete discussion of the entire process follows.

The first step in the process was to determine the average score for each IS graduate on each of the criteria, Y_i, $i = 1, 2,..., 9$. This score was determined for each of the criteria by averaging all the scores a graduate received from those judges who rated him on the criterion instrument previously discussed. From the previous regression run of Y_{11} on Y_1, Y_2,..., Y_9, the weights assigned to each of the predictor variables, Y_1, Y_2,..., Y_9, can be used to develop a prediction equation. (These weights were based on the single judgmental policy that existed for the IS department.)These average scores on each of the criteria along with the predetermined weight for the criteria were used to obtain a predicted score Y_{12} for each of the IS graduates. The predictive equation used was the following:

$$Y_{12} = 0.3119Y_1 + 0.1243Y_2 + 0.0000Y_3 + 0.2621Y_4 + 0.3292Y_5 + 0.3910Y_6$$
$$+ 0.2006Y_7 + 0.1200Y_8 + 0.1406Y_9 - 2.4473.$$

To determine the number of different policies existing among the IS faculty with regard to which characteristics they deemed most desirable in an IS doctoral graduate, a Type A normative JAN was completed on the data gathered from the 12 judges. The predictor variables used were the nine criterion variables listed in Section 4.4.3, namely Y_1, Y_2,..., Y_9. The criterion variable Y_{11} refers to the group in which each set of ratings was placed. The range of Y_{11} went from 1 to 7. Group 7 represented those sets of ratings each judge considered to be the most favorable in the 40 sets of ratings, and Group 1 represented those sets of ratings that each judge considered least favorable. Groups 2 through 6 lay in the continuum between the two extremes.

The regression flowchart in Figure 5.4 indicates, with respect to the existing judgmental policy of the 12 judges, that the affective variables, Y_4,

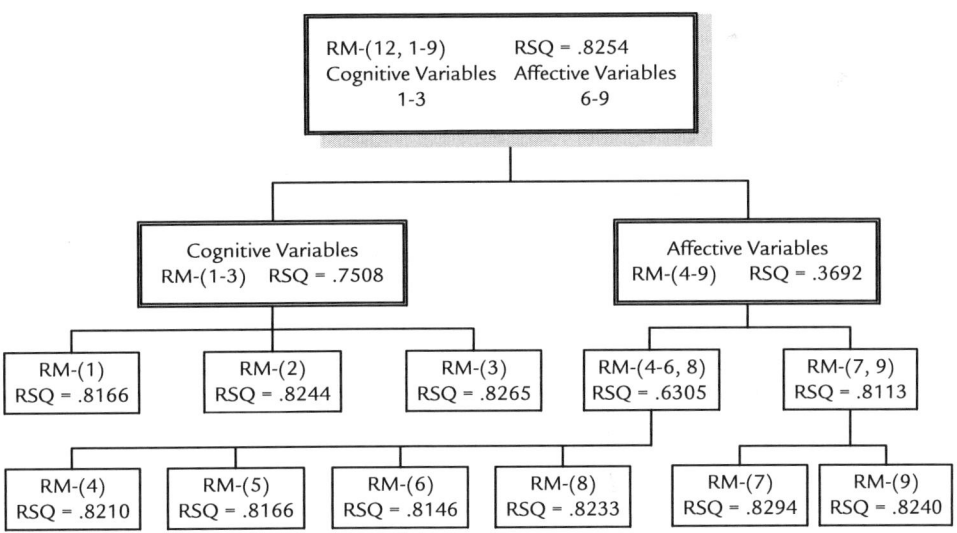

Note: Variables 1, 2, 3, ... , 9, 12
    12, 1-9 means a model using 12 as the criterion and 1-9 as predictors
    RM-( ) represents the restricted model formed by dropping variables ( ) out of the full model

FIGURE 5.4 Final Regression Chart for the Criterion Variables.

$Y_5,..., Y_9$, are more important than the cognitive variables, Y_1, Y_2, and Y_3. Of the affective variables, Y_4, Y_5, Y_6, and Y_8 played the biggest part in determining those doctoral candidates that the IS faculty considered outstanding or whom they would hire.

Next, a regression model was completed using the predicted score Y_{12} as the criterion and the variables X_2, X_3,..., X_{17}, which are the variables used to determine admittance to the program, as predictors. The regression flowchart in Figure 5.4 was developed through a factor analysis of the correlation matrix of predictors X_2, X_3,..., X_{17}. The correlation matrix is listed in Appendix B. The number of factors and the factor loadings suggested the grouping of subsets of variables as indicated on the flowchart. Variable X_7 was dropped from both of the cognitive subsets and individually because it loaded on both of these factors. In Figure 5.5, Y_{12} is denoted by variable 1.

An examination of the flowchart reveals that variables X_2, X_3, X_4, X_5, and X_9 when considered as a group have the greatest importance in determining admittance to the IS doctoral program based on the judgmental policy of the IS faculty as a single group. It should be noted that because of the

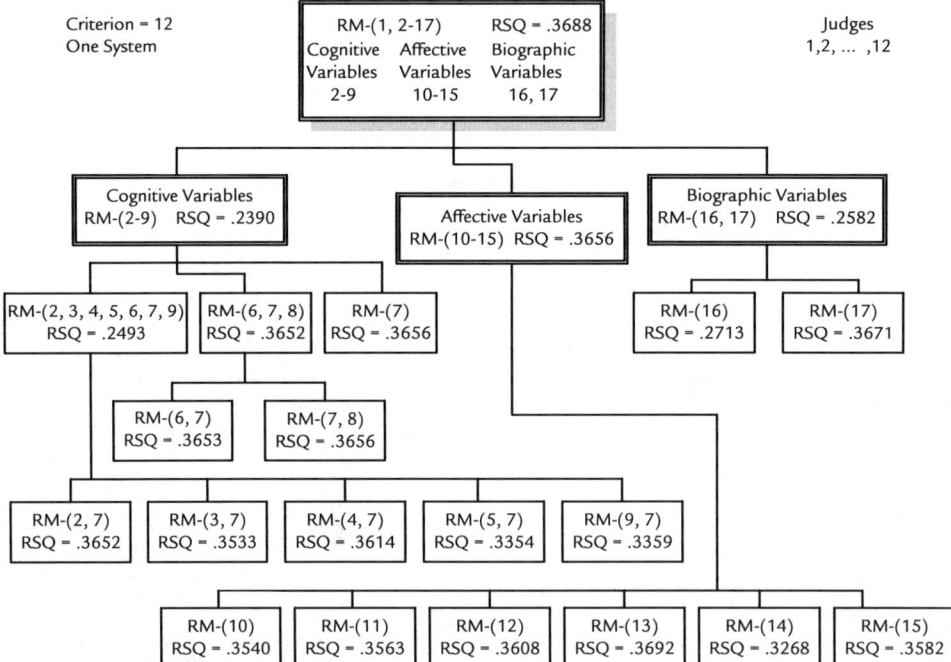

FIGURE 5.5 Final Regression Chart for the Predictor Variables.

multi-collinearities existing between these variables, there is virtually no R^2 drop at all when they are dropped out individually. The only other variable that gives a large drop in R^2 is variable X_{16}.

By using the methodology outline in the previous steps, the following results were obtained:

First, one judgmental policy existed, without arbitration, with regard to what constitutes an outstanding graduate.

Second, with regard to this policy, the IS faculty considered the affective variables more important than the cognitive variables. Of these affective variables, Y_4, Y_5, Y_6, and Y_8 were the most important.

Y_4 = Interpersonal relationships–faculty and supervisors

Y_5 = Interpersonal relationships–peers

Y_6 = Interpersonal relationships–with people he is serving (students, inmates, patients, etc.)

Y_8 = Personal characteristics (emotionally mature, friendly, enthusiastic)

Third, with regard to the judgmental policy of the IS faculty, variables X_2, X_3, X_4, X_5, X_9, and X_{16} are the most important in determining admittance to the IS doctoral program.

X_2 = GRE population score in social sciences

X_3 = GRE population score in humanities

X_4 = GRE population score in natural sciences

X_5 = GMAT verbal score

X_9 = Miller Analogies Test score

X_{16} = Age when admitted to the IS doctoral program

The R^2 value of 0.3688 for the full model is somewhat lower than expected. Possibly this is a result of the fact that the predictors, X_2, X_3,..., X_{17} are either cognitive or interest variables and do not relate to the affective criterion variables, Y_4, Y_5, Y_6, and Y_8 that the faculty deemed most important.

5.5 SUMMARY

The advantages of using JAN are numerous. If the number of distinct clusters, or policy groups, is not known in advance, which they normally wouldn't be, then multiple runs would also have to be run with multiple cluster levels. There are a number of options available for deciding on the correct number of clusters with JAN, use the F, or choose an R^2 drop level. The conclusions are obvious, with its high overall Jaccard mean across all conditions, JAN performs well in the task for which it was developed. And, by using a larger number of variables, ten or more, with a higher profile ratio, ten- or twenty-to-one, JAN's capture rate quickly increases.

The need to make only one run as opposed to multiple runs can lead to significant savings in both time and money. Not having to know in advance the actual number of distinct policy groups takes the guess work out of the process, thereby reducing the number of runs even further. Finally the program is easy to use and the output is easy to understand. The benefits of using the JAN technique in developing an objective decision policy are great. The process takes the individual raters subjectiveness into account and then develops a consensus policy which can be used to objectively score applicants on the criterion of interest. If used correctly, JAN is a very powerful tool.

5.6 EXERCISES

1. In a study of subscribers to two computing journals, a regression analysis was based on the data file to predict the value of the dependent variable TYPE (1 = major journal of the field, 0 = proceedings of national meeting) from the predictors ORGAN (professional organization affiliation), EDUC (degree held), INCOME (total household income), and INTEREST (research interests). The data file contained the values of these five variables for 225 individuals, and the partial chart in Figure 5.6 summarizes the results.

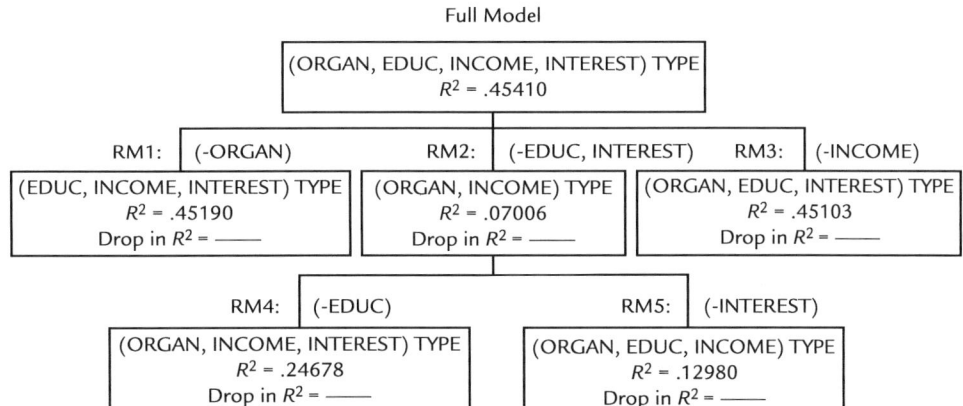

Full Model

FIGURE 5.6 Hypothetical Regression Analysis.

The partial printout in Table 5.9 corresponds to the full regression model.

Multr	0.67387	Anova	DF	F
Rsquare	0.45410	Regression	2	23.09671
Adjrsq	0.45308	Residual	222	Sign
Sterr	0.79004			.002

TABLE 5.9 ANOVA Regression Table.

a. Is the full model a "good" model from which to develop the regression equation? Why or why not?

b. Which predictors are significant in predicting TYPE? Which should be included in the final regression equation? Why?

2. The intercorrelation matrix in Table 5.10 relates overhead cost in a factory (Y) to units produced (X_1), direct labor cost (X_2), weight of output (X_3), and research and development costs (X_4).

	Y	X_1	X_2	X_3	X_4
Y	1.000	0.562	0.401	0.197	0.465
X_1	0.562	1.000	0.396	0.215	0.583
X_2	0.401	0.396	1.000	0.345	0.546
X_3	0.197	0.583	0.546	1.000	0.365
X_4	0.465	0.583	0.546	0.365	1.000

TABLE 5.10 Hypothetical Inter-correlation Matrix.

   a. Write out the equation R β_z = V.

   b. Find the normalized regression equation $Z_Y = \sum_{i=1}^{4} \beta_i z_{X_i}$ by the iterative Kelley-Salisbury technique.

   c. Find and interpret the value of R^2.

3. To aid in the selection of salesmen, the Jones Corporation administers two fifteen-minute tests. The first test is an achievement test that is designed to measure knowledge of leisure-time activities and current events. The second test gives a measure of the applicant's aggressiveness. The applicant is asked if he graduated from college; one (1) means that the applicant has graduated from college and zero (0) means that he/she has not.

Five Jones Corporation executives were given three profile scores for each of ten applicants. The executives were asked to rank the applicants from first choice (assigned number one) to last choice (assigned number ten). Table 5.11 shows the profile scores and the corresponding ranks.

	Profile Scores			Judgments				
ID	Test I	Test II	College	Judge 1	Judge 2	Judge 3	Judge 4	Judge 5
1	33	16	1	3	3	3	2	1
2	57	17	0	6	1	2	5	2
3	47	12	1	4	2	1	1	4
4	54	12	0	7	9	10	10	7
5	41	12	1	5	10	9	9	10
6	57	15	1	1	5	5	3	3
7	49	16	0	8	4	4	6	6
8	45	15	0	9	6	6	7	8
9	54	14	1	2	8	8	4	5
10	41	15	0	10	7	7	8	9

TABLE 5.11 Profile and Judgmental Scores.

Perform a JAN analysis to capture the optimal clustering of judges and then determine what each cluster of judges has for their judgmental policy. This analysis can be done either manually or through an implementation of JAN found in Appendix E.

4. Apply the JPC algorithm to the data in Table 5.2.

5. Both JAN and JPC are referred to as judgmental policy capturing methods. What is the difference between "policy capturing" and "judgment analysis"?

6. Try to replicate the JAN study in this chapter for a comparable application within an accessible organization. Additionally. Apply the JPC methodology recommended by Harvill, Lang, and McCord.

7. Using the data file in Problem 3, perform a Jancey clustering on the judges R^2 values. Are the final results comparable to the JAN results?

8. Is JAN an optimization method? Defend your answer.

9. What impact would the similarity between the policies have in a JAN study?

10. What are some of the factors that would impact JAN's ability to cluster?

11. Design a Monte Carlo study to validate the JAN method. Consider increasing the number of predictor variables, the number of cases to a profile, and the number of actual or expectant policies.

12. Design and implement JAN studies for the following: rating candidates applying for membership in an organization, rating TV shows for promotion in the job environment, etc. Let your imagination run wild and have fun.

FUZZY CLUSTERING MODELS AND APPLICATIONS

In This Chapter

6.1 INTRODUCTION

Fuzzy logic provides a means whereby imprecise or ambiguous data can be modeled. In general, humans do not think in terms of crisp numerical values such as, "The book is located 6.2 inches from the back, 2.7 inches from the left, on the third shelf from the base of the bookcase." This would probably be expressed more in terms such as, "The book is at the front left of the third shelf from the bottom." Fuzzy logic, like humans, allows for expression in linguistic terms instead of numeric values. Fuzzy systems map input variables to output variables by using linguistic rules instead of mathematical formulae. These concepts of fuzzy logic can then be applied to clustering methods.

In crisp logic, a value either has full membership or no membership, but in fuzzy logic, a value can have partial membership in different term sets. For example, take the concept of cold. How do we define a temperature that is cold? Few people could agree on one temperature that could be set as a threshold such that all temperatures below that would be cold. This would be an attempt at applying a crisp set to an ambiguous and imprecise problem.

Zadeh[1] defined fuzzy set as follows: A fuzzy set, A, on a universe of discourse, U, is characterized by a membership function $\mu_A(x)$ that takes values in the interval [0,1]. Various fuzzy sets can be used to define an attribute, such as *low*, *medium*, or *high*. An attribute with a certain value can also be a member of more than one set, such as 0.6 membership in term set *low* and 0.4 membership in term set *medium*, based on the output of the membership functions.

Fuzzy rules are used to map inputs to outputs by using these linguistic labels. Combined, fuzzy rules, input, and output form a fuzzy influence system. There are two basic classifications of fuzzy systems: the *Mamdani model* and the *Sugeno model*. An example of the Mamdani model would be a rule such as:

If x_i is *low* and x_2 is *high* then y_i is B.

In the Mamdani model, the emphasis is on using linguistic terms to describe the rule. In the Sugeno model, the output is obtained as a linear combination of the fuzzy inputs (Mitra and Hayashi).[2]

All of the clustering methods presented so far generate partitions, called *hard clusterings*. A partition has disjoint clusters, therefore, each pattern is a member of only one cluster. Fuzzy clusterings allow for a pattern to belong to more than one cluster, which is not a partition.

---

[1] Zadeh, L. A. (1965). Fuzzy Sets. *Information and Control, 8*, 338-353.

[2] Mitra, S. & Hayashi, Y. (2000). Neuro-Fuzzy Rule Generation: Survey in Soft Computing Framework. *IEEE Transactions on Neural Networks, 11*(3), 748-768.

[3] Jain, A. K., Murty, M. N., & Flynn, P. J. (1996). Data Clustering: A Review. [based on the chapter "Image Segmentation Using Clustering" in *Advances in Image Understanding: A Festschrift for Aerial Rosenfeld* (Bowyer, K. & Ahuja, N., eds.) (c) Computer Society Press.

Jain, Murty, and Flynn[3] provide the following fuzzy clustering algorithm:

---

(1) Select an initial fuzzy partition of the N objects into K clusters by selecting the $N \times K$ membership matrix U. An element μ_{ij} of this matrix represents the grade of membership of object x_i in cluster c_j. Typically, μ_{ij} [0,1].

(2) Using U, find the value of a fuzzy criterion function, e.g., a weighted squared error criterion function, associated with the corresponding partition. One possible fuzzy criterion function is

$$E^2(X,U) = \sum_{i=1}^{N} \sum_{k=1}^{K} \mu_{ij} \left\| x_i - c_k \right\|^2,$$

where $c_k = \sum_{i=1}^{N} \mu'_{ik} x_i$ and $\mu'_{ik} = \dfrac{\mu_{ik}}{\sum_{i=1}^{N} \mu_{ik}}$ is the kth fuzzy cluster center.

Reassign patterns to clusters to reduce this criterion function value and recompute U.

(3) Repeat step 2 until entries in U do not change significantly.

FIGURE 6.1 Fuzzy Clustering Algorithm.

Consider the set of points A = {(2,3), (3,2), (1,4), (3,3), (5,3)} and B = {(6,3), (7,2), (7,4), (8,3), (7,3)}. Additionally, assume the use of the Euclidean distance similarity measure, then potential neighborhoods for A and B could be disjoint as illustrated in Figure 6.2.

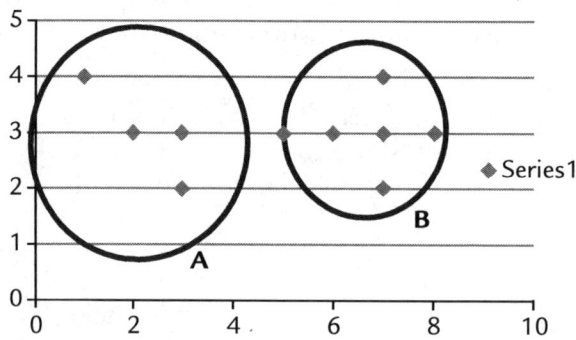

FIGURE 6.2 Traditional Non-fuzzy Neighborhoods.

Consider a different situation where A has a larger radius, as illustrated in Figure 6.3. Then the neighborhoods overlap.

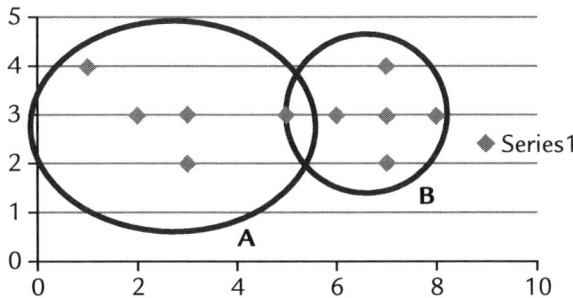

FIGURE 6.3 Fuzzy Neighborhoods.

In this case, (5,3) belongs to cluster A and cluster B. Because (5,3) is closer to the center of B, a radius approximately of 1, then (5,3) is to the center of A, a radius of approximately 3, we could define (5,3)'s grade of membership of A to be 0.33 and (5,3)'s grade of membership in B as 0.67. Larger membership values indicate higher confidence in the assignment of the pattern to the cluster.

In order to assign grade of memberships to every point, the points need to be labeled first.

Label	Point	
1	2	3
2	3	2
3	1	4
4	3	3
5	5	3
6	6	3
7	7	2
8	7	4
9	8	3
10	7	3

TABLE 6.1 Labeled Points in Euclidean Space.

Upon studying Table 6.1 the investigator might represent the fuzzy clusters Figure 6.4 using the grade of memberships as ordered pairs (pattern id, grade of membership in the fuzzy cluster being defined).

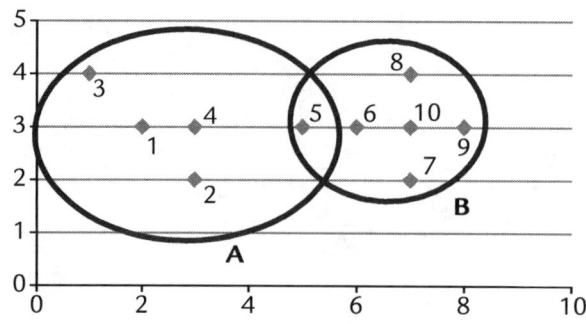

FIGURE 6.4 Labeled Points Fuzzy Neighborhoods.

Then the fuzzy clusters are A = {(3,0.05), (1,0.3), (4,0.3), (2,0.3), (5,0.05)} and B = {(8,0.2), (5,0.025), (6,0.3), (10,0.3), (9,0.025), (7,0.15)}. Because the grade of membership literally means "the degree of belonging to a specific cluster," then the sum of the grades of membership for all patterns in the cluster must equal one. Note that grade of membership is not a probability. Probability requires that a pattern belongs to only one cluster while grade of membership allows the pattern to belong to several or all clusters simultaneously. Again the grade of memberships in a cluster for this example were based on a ranking of the distance of the points from the center of the cluster.

Application of fuzzy set theory to clustering was introduced by Ruspini.[4] The interested reader should to consult Bezdek[5] and Zadeh[1] for material on fuzzy clustering. A fuzzy modification of the K means method called the fuzzy c-means (FCM) algorithm is by Bezdek.[5]

As one would suspect, the design of membership functions is the most important problem in fuzzy clustering; different choices include those based on similarity decomposition (Bezdek)[5] and centroids of clusters. A generalization of the FCM algorithm was proposed by Bezdek[5] through a family of objective functions. A fuzzy c-shell algorithm and an adaptive variant for detecting circular and elliptical boundaries was presented in Dave.[6]

[4] Ruspini, E. H. (1969). A new approach to clustering. *Information and Control, 15*, 22-32.

[5] Bezdek, J. C. (1981). *Recognition with Fuzzy Objective Function Algorithms.* New York: Plenum Press.

[6] Dave, R. N. (1992). Generalized fuzzy C-schells clustering and detection of circular and elliptic boundaries. *Pattern Recognition, 25*, 713-722.

6.2 THE MEMBERSHIP FUNCTION

The primary step for fuzzy clustering is to define a membership function based on similarity decompositions. One approach is presented in Backer.[7] Start with a given a set of patterns, $\{x_1, x_2,\ldots, x_n\}$, that are initially partitioned into clusters, $\{C_1, C_2,\ldots, C_K\}$. Assume that n_i is the number of patterns in C_i. Let $\delta(x,C_i)$ be the similarity between the pattern x and the cluster C_i. The cluster membership function is then defined as:

$$f_{C_i}(x) = P_i\delta(x,C_i)\Big/\sum_{k=1}^{K} P_i\delta(x,C_i),$$

where $P_i = n_i/n$. Then (1) $f_{C_k}(x) \geq 0$ and (2) $\sum_{k=1}^{K} f_{C_i}(x) = 1$, for all patterns. If $\delta(x,C_i)$ is a dissimilarity measure, like the Euclidean distance measure, then smaller, not larger, values of membership indicate higher confidence in the assignment of the pattern to the cluster.

For the clusters A and B in Figure 6.4, the current centroids are the following:

centroid for $A = \left(\dfrac{2+3+1+3+5}{5}, \dfrac{3+2+4+3+3}{5}\right) = (2.8,3)$ and centroid

for $B = \left(\dfrac{5+6+7+7+8+7}{6}, \dfrac{3+3+2+4+3+2}{6}\right) = (6.67,2.83)$.

Note that $\delta(x,C_i)$ is the Euclidean distance measure, not a similarity measure. One way to transform the Euclidean distance measure is to let $\delta(x,C_i)$ be the reciprocal of the Euclidean distance measure. Next, the grade membership value is found for every pattern and cluster combination.

Cluster A:

$$\delta(1,A) = 1\Big/\sqrt{(2.8-2)^2+(3-3)^2} = 1/0.2 = 5;$$
$$\delta(5,B) = 1\Big/\sqrt{(6.67-5)^2+(2.83-3)^2} = 1/1.68 = 0.60$$

$$\delta(2,A) = 1\Big/\sqrt{(2.8-3)^2+(3-2)^2} = 0.98;$$
$$\delta(6,B) = 1\Big/\sqrt{(6.67-6)^2+(2.83-3)^2} = 1.44$$

$$\delta(3,A) = 1\Big/\sqrt{(2.8-1)^2+(3-4)^2} = 0.98;$$
$$\delta(7,B) = 1\Big/\sqrt{(6.67-7)^2+(2.83-2)^2} = 1.12$$

---

[7] Backer, E. (1978). *Cluster Analysis by Optimal Decomposition of Induced Fuzzy Sets.* Delft, The Netherlands: Delft University Press.

$$\delta(4,A) = 1\Big/\sqrt{(2.8-3)^2 + (3-3)^2} = 0.98;$$
$$\delta(8,B) = 1\Big/\sqrt{(6.67-7)^2 + (2.83-4)^2} = 0.82$$

$$\delta(5,A) = 1\Big/\sqrt{(2.8-5)^2 + (3-3)^2} = 0.45;$$
$$\delta(9,B) = 1\Big/\sqrt{(6.67-8)^2 + (2.83-3)^2} = 0.75$$
$$\delta(10,B) = 1\Big/\sqrt{(6.67-3)^2 + (2.83-3)^2} = 0.27$$

and

$$\delta(1,B) = 1\Big/\sqrt{(6.67-2)^2 + (2.83-3)^2} = 0.21;$$
$$\delta(5,A) = 1\Big/\sqrt{(2.8-5)^2 + (3-3)^2} = 0.46$$

$$\delta(2,B) = 1\Big/\sqrt{(6.67-3)^2 + (2.83-2)^2} = 0.27;$$
$$\delta(6,A) = 1\Big/\sqrt{(2.8-6)^2 + (3-3)^2} = 0.31$$

$$\delta(3,B) = 1\Big/\sqrt{(6.67-1)^2 + (2.83-4)^2} = 0.17;$$
$$\delta(7,A) = 1\Big/\sqrt{(2.8-7)^2 + (3-2)^2} = 0.23$$

$$\delta(4,B) = 1\Big/\sqrt{(6.67-3)^2 + (2.83-3)^2} = 0.27;$$
$$\delta(8,A) = 1\Big/\sqrt{(2.8-7)^2 + (3-4)^2} = 0.30$$

$$\delta(5,B) = 1\Big/\sqrt{(6.67-5)^2 + (2.83-3)^2} = 0.60;$$
$$\delta(9,A) = 1\Big/\sqrt{(2.8-8)^2 + (3-3)^2} = 0.19$$
$$\delta(10,A) = 1\Big/\sqrt{(2.8-3)^2 + (3-3)^2} = 5$$

then $P_A = \dfrac{5}{10} = 0.5$ and $P_B = \dfrac{6}{10} = 0.6$ and

$$f_A(1) = \frac{P_A\delta(1,A)}{\left[P_A\delta(1,A) + P_B\delta(1,B)\right]} = \frac{(0.5)(5)}{[(0.5)(5) + (0.6)(.21)]} = 0.17$$

$$f_A(2) = \frac{P_A\delta(2,A)}{\left[P_A\delta(2,A) + P_B\delta(2,B)\right]} = \frac{(0.5)(0.98)}{[(0.5)(0.98) + (0.6)(0.21)]} = 0.62$$

$$f_A(3) = \frac{P_A\delta(3,A)}{\left[P_A\delta(3,A) + P_B\delta(3,B)\right]} = \frac{(0.5)(0.98)}{[(0.5)(0.98) + (0.6)(0.17)]} = 0.83$$

$$f_A(4) = \frac{P_A\delta(4,A)}{\left[P_A\delta(4,A) + P_B\delta(4,B)\right]} = \frac{(0.5)(0.98)}{[(0.5)(0.98) + (0.6)(0.27)]} = 0.75$$

$$f_A(5) = \frac{P_A\delta(5,A)}{\left[P_A\delta(5,A) + P_B\delta(5,B)\right]} = \frac{(0.5)(0.45)}{[(0.5)(0.45) + (0.6)(0.6)]} = 0.38$$

then

$$f_B(5) = \frac{P_B\delta(5,B)}{\left[P_A\delta(5,A) + P_B\delta(5,B)\right]} = \frac{(0.6)(0.6)}{[(0.5)(0.45) + (0.6)(0.6)]} = 0.62$$

note that $f_B(5) + f_A(5) = 0.62 + 0.38 = 1.0$.

therefore $\quad f_B(1) = 0.83, \; f_B(2) = 0.38, \; f_B(3) = 0.17, \; f_B(4) = 0.25$

and $\quad\quad f_B(6) = 0.15, \; f_B(7) = 0.14, \; f_B(8) = 0.23, \; f_B(9) = 0.17, \; f_B(10) = 0.94$

$\quad\quad\quad\quad f_A(6) = 0.85, \; f_A(7) = 0.86, \; f_A(8) = 0.77, \; f_A(9) = 0.83, \; f_A(10) = 0.06.$

6.3 INITIAL CONFIGURATION

Using the membership functional values for A and B serving as an initial configuration let U be the membership matrix, as illustrated in Table 6.2:

Point	Membership in	
	A	B
1	0.17	0.83
2	0.62	0.38
3	0.83	0.17
4	0.75	0.25
5	0.38	0.62
6	0.85	0.15
7	0.86	0.14
8	0.77	0.23
9	0.83	0.17
10	0.06	0.94

$U =$

TABLE 6.2 Membership Matrix.

In iteration one: for each pattern, or point, compute the weighted square error criterion function value with respect to the new fuzzy cluster centers,

then

$$C_A = ((0.17)(2.8,3) + (0.62) * (2.8,3) + (0.83) * (2.8,3) + (0.75) * (2.8,3) + (0.38) * (2.8,3))$$

$$= (0.48,0.51) + (1.74,1.86) + (2.32,2.49) + (2.1,2.25) + (1.06,1.14)$$

$$= (7.7,5.25)$$

and

$$(0.17 + 0.62 + 0.83 + 0.75 + 0.38) = 2.75$$

$$\text{the } C_A = (7.7/2.75, 5.25/2.75) = (2.8,1.91)$$

Similarly $C_B = (6.89,2.98)$.

6.4 MERGING OF CLUSTERS

Start with clustering $\{A,B\}$. Consider:

$$E^2(1,U_1) = (0.17 * [1/\{(2-2.8)^2 + (3-2.35)^2\}^{1/2}] + 0.93 * [1/\{(2-6.46)^2 + (3-2.83)^2\}^{1/2}] = 0.35$$

$$E^2(2,U_2) = (0.62 * [1/\{(3-2.8)^2 + (2-2.35)^2\}^{1/2}] + 0.38 * [1/\{(3-6.46)^2 + (2-2.83)^2\}^{1/2}] = 1.64$$

$$E^2(3,U_3) = (0.83 * [1/\{(1-2.8)^2 + (4-2.35)^2\}^{1/2}] + 0.17 * [1/\{(1-6.46)^2 + (4-2.83)^2\}^{1/2}] = 0.37$$

$$E^2(4,U_4) = (0.75 * [1/\{(3-2.8)^2 + (3-2.35)^2\}^{1/2}] + 0.25 * [1/\{(3-6.46)^2 + (3-2.83)^2\}^{1/2}] = 0.38$$

$$E^2(5,U_5) = (0.38 * [1/\{(5-2.8)^2 + (3-2.35)^2\}^{1/2}] + 0.62 * [1/\{(5-6.46)^2 + (3-2.83)^2\}^{1/2}] = 0.54$$

$$E^2(6,U_6) = (0.85 * [1/\{(6-2.8)^2 + (3-2.35)^2\}^{1/2}] + 0.15 * [1/\{(6-6.46)^2 + (3-2.83)^2\}^{1/2}] = 0.57$$

$$E^2(7, U_7) = (0.86 * [1/\{(7 - 2.8)^2 + (2 - 2.35)^2\}^{1/2}] + 0.14 * [1/\{(7 - 6.46)^2 + (2 - 2.83)^2\}^{1/2}] = 0.35$$

$$E^2(8, U_8) = (0.77 * [1/\{(7 - 2.8)^2 + (4 - 2.35)^2\}^{1/2}] + 0.23 * [1/\{(7 - 6.46)^2 + (4 - 2.83)^2\}^{1/2}] = 0.35$$

$$E^2(9, U_9) = (0.83 * [1/\{(8 - 2.8)^2 + (3 - 2.35)^2\}^{1/2}] + 0.17 * [1/\{(8 - 6.46)^2 + (3 - 2.83)^2\}^{1/2}] = 0.43$$

$$E^2(10, U_{10}) = (0.06 * [1/\{(7 - 2.8)^2 + (3 - 2.35)^2\}^{1/2}] + 0.94 * [1/\{(7 - 6.46)^2 + (3 - 2.83)^2\}^{1/2}] = 1.68$$

$$E^2(x, U) = \sum_{i=1}^{N} (x_i, U_i) = 6.66 \text{ is the weighted squared error for the present}$$
clustering.

If we merge the clusters into a single cluster, then a single center must be found for all the patterns.

Point	Membership in	
	A	B
1	0.17	0.83
2	0.62	0.38
3	0.83	0.17
4	0.75	0.25
5	0.38	0.62
$U=$ 6	0.85	0.15
7	0.86	0.14
8	0.77	0.23
9	0.83	0.17
10	0.06	0.94

TABLE 6.3 Two Cluster Membership Matrix.

The membership matrix in Table 6.3 needs to be made into a single column. One rule that would apply is to capture the maximum value in each row of the existing matrix U, because this grade of membership is acceptable for A∪B:

Point	Membership in
1	0.83
2	0.62
3	0.83
4	0.75
5	0.62
6	0.85
7	0.86
8	0.77
9	0.83
10	0.94

$U =$

TABLE 6.3 Two Cluster Membership Matrix.

then the single centroid $C = [(.83)(2,3) + (.62)(3,2) + (.83)(1,4) + (.75)(3,3) + (.62)(5,3) + (.85)(6,3) + (.86)(7,2) + (.77)(7,4) + (.83)(8,3) + (.94)(7,3)]$ $[1/(.83 + .62 + \ldots + .83 + .94)]$

$$C = (39.43/7.9, 21.33/7.9) = (4.99, 2.7)$$

which allows for the computation of the weighted square:

$$E^2(1,U) = (0.83) * [1/\{(2 - 4.99)^2 + (3 - 3.17)^2\}^{1/2}] = 0.28$$

$$E^2(2,U) = (0.62) * [1/\{(3 - 4.99)^2 + (2 - 3.17)^2\}^{1/2}] = 0.29$$

$$E^2(3,U) = (0.83) * [1/\{(3 - 4.99)^2 + (4 - 3.17)^2\}^{1/2}] = 0.38$$

$$E^2(4,U) = (0.75) * [1/\{(3 - 4.99)^2 + (3 - 3.17)^2\}^{1/2}] = 0.38$$

$$E^2(5,U) = (0.62) * [1/\{(5 - 4.99)^2 + (3 - 3.17)^2\}^{1/2}] = 0.75$$

$$E^2(6,U) = (0.85) * [1/\{(6 - 4.99)^2 + (3 - 3.17)^2\}^{1/2}] = 0.83$$

$$E^2(7,U) = (0.86) * [1/\{(7 - 4.99)^2 + (2 - 3.17)^2\}^{1/2}] = 0.37$$

$$E^2(8,U) = (0.77) * [1/\{(7 - 4.99)^2 + (4 - 3.17)^2\}^{1/2}] = 0.35$$

$$E^2(9,U) = (0.83) * [1/\{(8 - 4.99)^2 + (3 - 3.17)^2\}^{1/2}] = 0.28$$

$$E^2(10,U) = (0.94) * [1/\{(7 - 4.99)^2 + (3 - 3.17)^2\}^{1/2}] = 0.47$$

$$E^2(x,U) = \sum_{i=1}^{N} (x_i, U) = 4.38 \text{ is the weighted squared error for the}$$
present clustering.

In this case, the weighted squared error for the single clustering, $A \cup B$, has a smaller weighted squared error value than the same value found for the clustering $\{A,B\}$. This result might infer that the single cluster is the clustering to best represent the data, however, at least a complete clustering method run on the data needs to be completed before determining the final clustering.

6.5 FUNDAMENTALS OF FUZZY CLUSTERING

Given a data set X, a crisp clustering partitions X into clusters $\{C_i \mid 1 \leq i \leq k, k = \textit{number of clusters}\}$ with the properties:

$$\bigcup_{i=1}^{K} C_i = X$$

$$C_i \cap C_j = \varnothing, \text{ for } 1 \leq i, j \leq K \text{ and } i \neq j$$

$$\varnothing \subseteq C_i, 1 \leq i \leq K.$$

For such a partition, let U be the membership matrix where each row of U contains the membership function values $f_i(C_k)$, or the membership function value of data point i (row i in X) being a member of cluster C_k. For a crisp partitioning:

$$f_i(C_k) \in \{0,1\}, 1 \leq k \leq K, 1 \leq i \leq N, \text{ where } N \text{ is the number of data points in}$$
$$X, \text{ and there are } K \text{ clusters}$$

$$\sum_{i=1}^{K} f_i(C_k), 1 \leq i \leq N$$

$$0 < \sum_{i=1}^{N} f_i(C_k) < N, 1 \leq k \leq K.$$

For a fuzzy partitioning, we simply relax the first condition:

$$f_i(C_k) \in [0,1], 1 \leq k \leq \mathrm{K}, 1 \leq i \leq N, \text{ where } N \text{ is the number of data points in}$$
$$X, \text{ and there are } K \text{ clusters}$$

$$\sum_{i=1}^{K} f_i(C_k), 1 \leq i \leq N$$

$$0 < \sum_{i=1}^{N} f_i(C_k) < N, 1 \leq k \leq K.$$

Consider the example presented in the last section given the following data set:

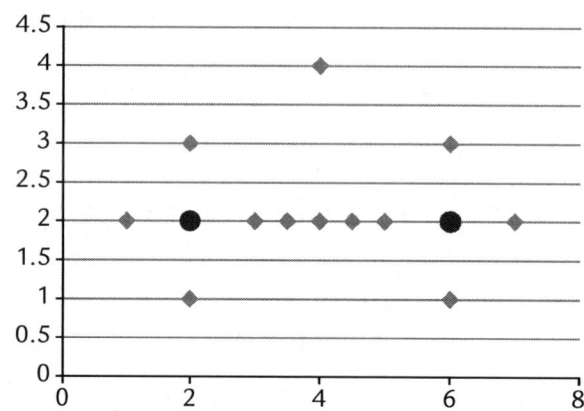

FIGURE 6.5 Data Set with Centroids (2,2) and (6,2) .

The similarity measures are then:

δ(x,A)	δ(x,B)
1.00	0.20
1.00	0.45
1.00	0.24
1.00	0.33
0.67	0.40
0.50	0.50
0.40	0.67
0.33	1.00
0.24	1.00
0.45	1.00
0.20	1.00
0.35	0.35

Now provide a new definition for $f_X(i)$ which meets the conditions for a fuzzy partitioning:

$$f_A(i) = \frac{\delta(i,A)}{\delta(i,A) + \delta(i,B)} \quad \text{and} \quad f_B(i) = \frac{\delta(i,B)}{\delta(i,A) + \delta(i,B)}.$$

The membership values for the fuzzy partitioning become the following:

$f_A(x)$	$f_B(x)$
0.83	0.17
0.69	0.31
0.81	0.19
0.75	0.25
0.67	0.33
0.50	0.50
0.33	0.67
0.25	0.75
0.25	0.75
0.31	0.69
0.17	0.83
0.50	0.50

Notice that data points (4,2) and (4,4) belong to both clusters A and B. The fact that the boundary point (4,2) has a membership value of 0.50 in both clusters correctly represents its position in the middle position between the two clusters. But even though (4,4) also has a membership value of 0.50, it is actually a greater distance from the centroids of (2,2) and (6,2) which means the point (4,4) has a smaller degree of similarity to the clusters A and B than the point (4,2). This type of situation can occur in a fuzzy partitioning. To overcome this situation possibilistic functions are utilized:

$f_i(C_k) \in [0,1]$, $1 \le k \le K$, $1 \le i \le N$, where N is the number of data points in X, and there are K clusters

there exists an i, $1 \le i \le N$, where $f_i(C_k) > 0$, for $1 \le k \le K$

$$0 < \sum_{i=1}^{N} f_i(C_K) < N, 1 \le k \le K.$$

6.6 FUZZY C-MEANS CLUSTERING

Most fuzzy clustering algorithms are objective function based: They determine an optimal classification by minimizing an objective function. In objective function based clustering usually each cluster is represented by a cluster prototype. This prototype consists of a cluster center, whose name already indicates its meaning, and maybe some additional information about the size and the shape of the cluster. The cluster center is an instantiation of the attributes used to describe the domain. However, the cluster center is computed by the clustering algorithm and may or may not appear in the data set. The size and shape parameters determine the extension of the cluster in different directions of the underlying domain.

The degrees of membership to which a given data point belongs to the different clusters are computed from the distances of the data point to the cluster centers with respect to the size and the shape of the cluster as stated by the additional prototype information. The closer a data point lies to the center of a cluster, the higher its degree of membership to this cluster. Hence, the problem is to minimize the distances of the data points to the cluster centers, because, of course, we want to maximize the degrees of membership.

Several fuzzy clustering algorithms can be distinguished depending on the additional size and shape information contained in the cluster prototypes, the way in which the distances are determined, and the restrictions that are placed on the membership degrees. Distinction is made, however, between probabilistic and possibilistic clustering, which use different sets of constraints for the membership degrees.

For each datum in a probabilistic cluster analysis, a probability distribution over the clusters is determined that specifies the probability with which a datum is assigned to a cluster. These techniques are also called fuzzy clustering algorithms if the probabilities are interpreted as degrees of membership. Possibilistic cluster analysis techniques are pure fuzzy clustering algorithms. Degrees of membership or possibility indicate to what extent a datum belongs to the clusters. Possibilistic cluster analysis drops the probabilistic constraint that the sum of memberships of each datum to all clusters is equal to one. Krishnapuram and Keller[8] emphasize that probabilistic clustering is primarily a partitioning algorithm, whereas possibilistic clustering is a rather mode-seeking technique, aimed at finding meaningful clusters.

---

[8] Krishnapuram, R. & Keller, J. (1996). The Possibilistic C-Means Algorithm: Insights and Recommendations. *IEEE Trans. Fuzzzy Systems, 4*, 385-393.

In review, consider the K-means clustering method. The algorithm is composed of the following steps:

1. *Start with K board games that are amongst the board games that are being clustered. These board games represent the initial group centroids.*

2. *Assign each board game to the group that has the closest centroid.*

3. *When all objects have been assigned, recalculate the positions of the K centroids.*

Repeat Steps 2 and 3 until the centroids no longer move. This produces a separation of the objects into groups from which the metric to be minimized can be calculated.

The K-means algorithm has problems when clusters are of differing sizes, densities, and nonglobular shapes. Additionally, the K-means algorithm encounters problems with outliers and empty clusters. In research, almost every aspect of K-means has been modified including: distance measures, centroid and objective function definitions, the overall process and efficiency enhancements, especially in initialization. New distance measures have been utilized such as the cosine measure, and the Jaccard measure. Bregman[9] divergence measures allow a K-means type algorithm to apply to many distance measures.

In the fuzzy c-means clustering, an object belongs to all clusters with some weight and the sum of the weights is 1. An excellent source of information on this algorithm is by Bezdek.[10] Harmonic K-means uses the harmonic mean instead of standard mean. A general reference to various flavors of K-means clustering is available online, *Introduction to Data Mining* by Tan, Steinbach, and Kumar[11] Addision-Wesley, 2005 at *http://www-users.cs.umn.edu/~kumar/dmbook/index.php*. Another good resource is by Han and Kamber[12] also online at *http://www-sal.cs.uiuc.edu/~hanj/bk2*. CLUTO clustering software is available at *http://glaros.dtc.umn.edu/gkhome/views/clutorences*.

---

[9] Banerjee, A., Merugu, S., Dhillon, I., & Ghosh, J. (2005). Clustering with Bregman Divergences. *Journal of Machine Learning Research*.

[10] Bezdek, J. C. (1973). *Fuzzy Mathematics in Pattern Classification*, PhD Thesis. Ithaca, NY: Cornell University.

[11] Tan, P-N, Steinbach, M., & Kumar, V. (2005). *Introduction to Data Mining*. Addison Wesley.

[12] Han, J. & Kamber, M. (2006). *Data Mining: Concepts and Techniques, 2nd Edition*. Morgan Kauffman.

The fuzzy c-means clustering is simply the following modification of the K-means clustering:

Fuzzy C-Means Clustering

This algorithm is based upon iterative optimization of the objective function, with update of membership and cluster centers.

- This is based upon an initial membership matrix for each item in a cluster.

- Center of clusters are calculated based upon the membership function.

- Once the centers are determined, the membership matrix is updated.

When the difference between two sequential membership matrixes is less than the initial termination criterion, the algorithm is stopped. Otherwise Steps 2 and 3 are repeated.

Consider the following data set illustrated in Table 6.5 and Figure 6.6.

X	Y	Pt ID
0	4	1
0	3	2
1	5	3
2	4	4
3	3	5
2	2	6
2	1	7
1	0	8
5	5	9
6	5	10
7	6	11
5	3	12
7	3	13
6	2	14
6	1	15
8	1	16

TABLE 6.5 Fuzzy C-Means Test Data.

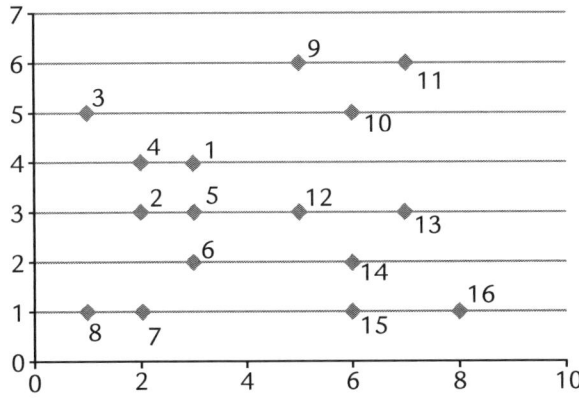

FIGURE 6.6 Scatter Gram for Fuzzy C-Means Test Data.

Assume two centroids: C1 = (2,3) and C2 = (7,3). Let the inverse of 2-dimensional Euclidean distance be the measure of similarity. Then the first iteration of the fuzzy c-means clustering algorithm generates:

Step 1: Choose an initial matrix U:

Pt 1	$u(x,1)$	$u(x,2)$
1	0.6	0.4
2	0.6	0.4
3	0.9	0.1
4	0.9	0.1
5	0.9	0.1
6	0.9	0.1
7	0.65	0.35
8	0.55	0.45
9	0.35	0.65
10	0.3	0.7
11	0.3	0.7
12	0.1	0.9
13	0.1	0.9
14	0.1	0.9
15	0.3	0.7
16	0.3	0.7

Step 2: Compute the new centroids

$$\sum_{i=1}^{N} u_{i1}\vec{v}_i = 20.7 \text{ and } \sum_{i=1}^{N} u_{i1}\vec{v}_i = 23.9$$

$$\sum_{i=1}^{N} u_{i1} = 7.85$$

then $C1 = (2.637, 3.045)$

Similarly, $C2 = (4.945, 2.957)$

where Cluster $1 = \{1,2,3,4,5,6,7\}$ and Cluster $2 = \{8,9,10,11,12,13,14,15,16\}$

Step 3: Update the membership matrix U

$$d_{11} = \sqrt{(0-2.637)^2 + (4-3.045)^2} \text{ and } \delta(1,1) = 1/d_{11} =$$

$$d_{12} = \sqrt{(0-4.945)^2 + (4-2.957)^2} \text{ and } \delta(1,2) = 1/d_{12} =$$

The new fuzzy weights are found using

$$u_{12}^{(1)} = \frac{u_{12}\delta(1,1)}{[u_{11}\delta(1,1) + u_{12}\delta(1,2)]} \text{ and } u_{11}^{(1)} = \frac{u_{11}\delta(1,1)}{[u_{11}\delta(1,1) + u_{12}\delta(1,2)]}$$

0.730	0.270
0.738	0.262
0.940	0.060
0.961	0.039
0.980	0.020
0.958	0.042
0.754	0.246
0.635	0.365
0.264	0.736
0.202	0.798
0.230	0.770
0.003	0.997
0.050	0.950
0.043	0.957
0.195	0.805
0.213	0.787

Step 4: Let $\epsilon = 0.001$

if $\left\| U^{(k+1)} - U^{(k)} \right\| < \epsilon$, where $U^{(k)}$ is the kth iteration U matrix, then Stop else start again at Step 2.

Iteration 2

Step 2 new centroids $C1 = (2.216, 2.995)$ and $C2 = (5.540, 7.041)$
with matrix U

0.890	0.110
0.904	0.096
0.974	0.026
0.990	0.010
0.988	0.012
0.991	0.009
0.880	0.120
0.772	0.228
0.163	0.837
0.099	0.901
0.140	0.860
0.001	0.999
0.015	0.985
0.013	0.987
0.098	0.902
0.116	0.884

Iteration 3

new centroids: $C1 = (1.8088, 2.9908)$ and $C2 = (5.929, 3.130)$
with a new matrix U

0.969	0.031
0.978	0.022
0.990	0.010
0.997	0.003
0.992	0.008
0.998	0.002
0.954	0.046
0.890	0.110
0.090	0.910
0.040	0.960
0.072	0.928
0.000	1.000
0.003	0.997
0.003	0.997
0.046	0.954
0.054	0.946

Iteration 7 Algorithm stops.

new centroids $C1 = (1.414, 2.768)$ and $C2 = (6.229, 3.237)$

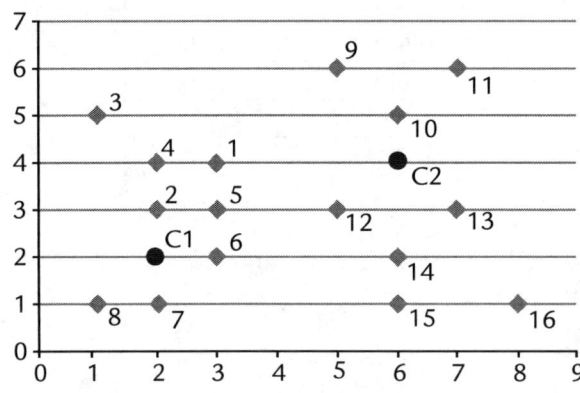

FIGURE 6.7 Final Scattergram.

with cluster membership matrix

0.99926	0.00074
0.99977	0.00023
0.99927	0.00073
0.99990	0.00010
0.99889	0.00111
0.99998	0.00002
0.99740	0.00260
0.99075	0.00925
0.01233	0.98767
0.00181	0.99819
0.00680	0.99320
0.00001	0.99999
0.00001	0.99999
0.00006	0.99994
0.00438	0.99562
0.00412	0.99588

A fuzzy partition of a data set X is one that characterizes the membership of each data point in all the clusters by a membership function, which is in [0,1]. A hard partition, such as a K-means clustering, is a special case of fuzzy partitions, where each data point belongs to one and only one cluster. Hard partitions have difficulty classifying outliers. However, fuzzy partitions resolve this difficulty.

Fuzzy clustering algorithms can generate hard partitions. Hard clustering algorithms cannot determine fuzzy partitions. Fuzzy partitions are an extension of hard partitions, because each data point in a fuzzy partition is assigned to a single cluster but is allowed to be in partial membership in several fuzzy clusters.

6.7 INDUCED FUZZINESS

A normal set of heights for tall men might be expressed as Tall = $\{x \mid x > 6 \text{ ft.}\}$. A fuzzy set extension for this set would be defined as Tall = $\{x, \mu_A(x) \mid x \in X\}$, where $\mu_A(x)$ is called the membership function of x in A. The membership function maps each element of X to a membership value between 0 and 1. To "fuzzify" data in a study, what is needed is a toolbox of commonly applicable membership functions. Some widely used membership functions include:

- piecewise linear functions,
- the Gaussian distribution function,
- the sigmoid curve, and
- quadratic and cubic polynomial curves.

These membership functions apply, when data is fuzzified, instead of the [0,1] step function:

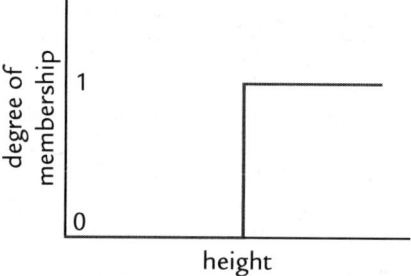

FIGURE 6.8 Crisp Membership Function.

The triangular and trapezoidal membership curves are useful for determining fuzzy membership values which are easily implemented.

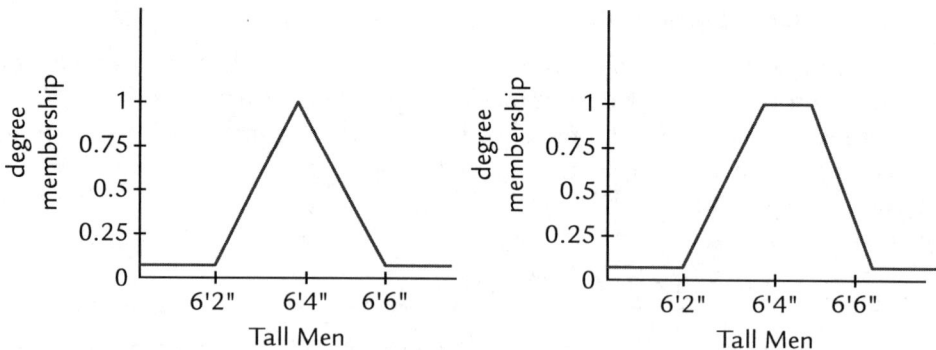

FIGURE 6.9 Triangular and Trapezoidal Membership Functions.

The formulas for these membership functions are shown below:

(a) Triangular

$$f(x; a, b, c) = \begin{cases} 0, & x \le 0 \\ 1 - \dfrac{a-x}{b-a}, & a \le b \\ \dfrac{c-x}{c-b}, & b \le x \le c \\ 0, & x \ge c \end{cases}$$

(b) Trapezoidal

$$f(x; a, b, c, d) = \begin{cases} 0, & x \le a \\ 1 - \dfrac{x-a}{b-a}, & a \le x \le b \\ 1, & b \le x \le c \\ \dfrac{d-x}{d-c}, & c \le d \\ 0, & d \ge x \end{cases}$$

The bell membership function, or the traditional normal distribution, offers smoothness coupled with a concise notation.

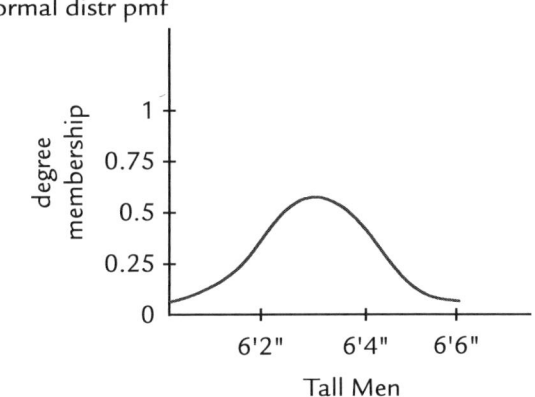

FIGURE 6.10 Bell, or Gaussian, Membership Function.

The formula for the Gaussian membership function is:

$$f(x: \sigma, c) = e^{\frac{-(x-c)^2}{2\sigma^2}}.$$

Another commonly used membership function, widely used in learning theory as a learning curve, is the sigmoidal curve. This is actually the cumulative distribution function for the normal probability function.

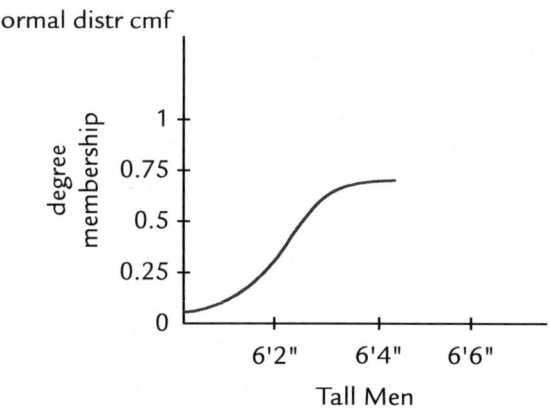

FIGURE 6.11 Sigmoid Membership Function.

The membership formula for the sigmoid curve is:

$$S(x; a, b, c) = \begin{cases} 0, & x < a \\ \dfrac{2(x-a)^2}{(c-a)^2}, & a \leq x \leq b \\ 1 - \dfrac{2(x-c)^2}{(c-a)^2}, & b \leq x \leq c \\ 1, & x > c \end{cases}.$$

Illustrations of each of these functions are given in Figure 6.12 based upon a minimum of 0 and a maximum of 1.0. Note that any number of membership functions may be used to express the values of an attribute, such as the five used in the illustration that correspond to very low, low, medium, high, and very high.

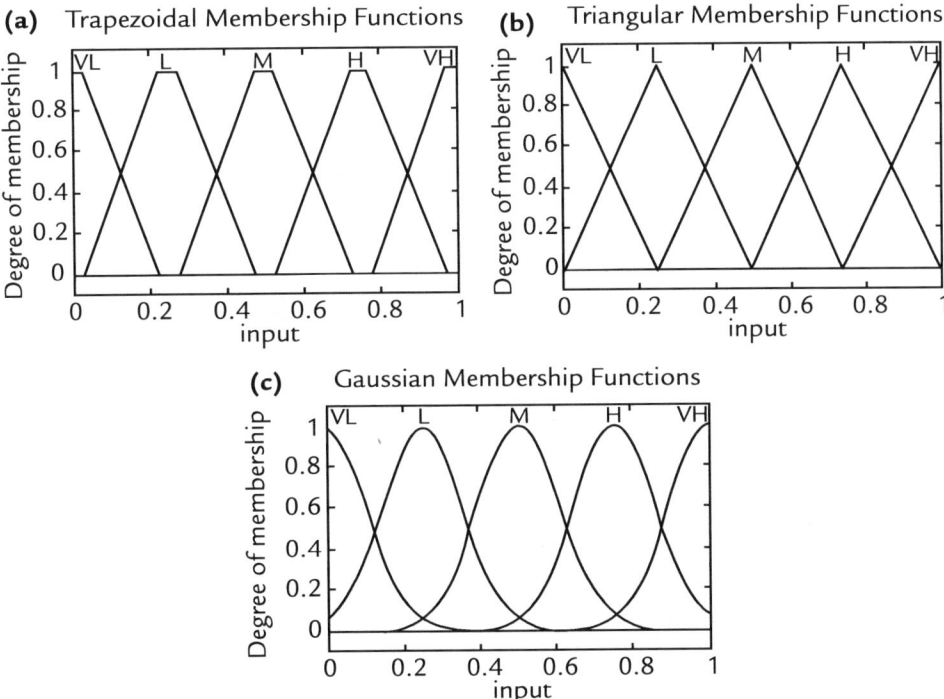

FIGURE 6.12 Examples of Fuzzy Membership Sets Using Five Functions: (a) Trapezoidal, (b) Triangular, and (c) Gaussian.

If numeric attributes are present, make the data fuzzy before continuing. Each numeric attribute separately should undergo fuzzy c-means clustering (Hoppner, Klawonn, Kruse, and Runkler)[13] in order to cluster each continuous value into one of the preset number of clusters that are represented as nominal values.

6.8 SUMMARY

Dr. Lotfi A. Zadeh initiated fuzzy logic in 1965. Fuzzy logic is a multivalued logic that allows intermediate values to be defined between conventional evaluations like true/false, yes/no, high/low, etc. Human language rules are the foundation for building a fuzzy logic system. These vague and

---

[13] Hoppner, F., Klawonn, F., Kruse, R., & Runkler, T. (1999). The fuzzy c-means algorithm. *Fuzzy Cluster Analysis: Methods for Classification, Data Analysis and Image Recognition.* West Sussex, England: John Wiley and Sons, Ltd., pp. 37-43.

ambiguous rules are then transformed in mathematical equivalents. Human language rules are often not only vague and ambiguous but can be imprecise or full of noise and even missing altogether. Fuzzy systems are better representations of worldly behavior than their crisp system counterparts.

Fuzzy systems are especially applicable for nonlinear processes. They are helpful for very complex or highly nonlinear processes. On the other hand, sometimes results are unexpected and hard to debug in fuzzy systems. Additionally, as illustrated by the methods in this chapter, fuzzification can be computationally complicated. Due to this fact, crisp method is preferable if it yields a satisfying result.

In classical K-means procedure, each data point is assumed to be in exactly one cluster. In fuzzy K-means clustering, we can relax this condition and assume that each sample x_j has some graded or "fuzzy" membership in a cluster.

6.9 EXERCISES

1. Given the following points in 2 dimensional Euclidean space find (a) crisp neighborhoods and (b) fuzzy neighborhoods for clusterings containing only 2 clusters:

5	4.5
7	4
9	4
4	3.5
9	4.5
4	4.5
10	4
5	3.5
10	4.5
5	4
6	4
8	4
4	4
9	3.5
10	3.5

TABLE 6.6 Dimensional Euclidean Space Data Set.

2. Find the centroids for the neighborhoods for Problem 1.

3. Using the reciprocal of Euclidean distance as a similarity measure find the membership values for all the points in Problem 1.

4. Using the membership functional values found in Problem 3, find the new membership values at the end of one iteration of applying the weighted square error criterion function.

5. Merge the two neighborhoods in Problem 4 into a single neighborhood and find the new centroid.

6. Given the following data set:

X	Y	Pt. ID
3	4	1
2	3	2
1	5	3
2	4	4
3	3	5
3	2	6
2	1	7
1	1	8
5	6	9
6	5	10
7	6	11
5	3	12
7	3	13
6	2	14
6	1	15
8	1	16

TABLE 6.7 Second 2-Dimensional Euclidean Space Data Set.

perform a fuzzy c-means clustering on this data set with initial centroids $c1 = (2,2)$ and $c2 = (6,4)$.

7. Induce fuzziness on the following data set for software development using the:

   A. Triangular membership function,

   B. Trapezoidal membership function,

   C. Gaussian membership function, and

   D. Sigmoid membership function.

Id	Size	Effort	Duration
1	562	1062	14
2	647	7871	16
3	130	845	5
4	254	2330	8
5	1056	21272	16
6	383	4224	12
7	345	2826	12
8	209	7320	27
9	366	9125	24
10	1181	11900	54
11	181	4300	13
12	739	4150	21
13	108	900	7
14	48	583	10
15	249	2565	19
16	371	4047	11
17	211	1520	13
18	1849	25910	32
19	2482	37286	38
20	434	15052	40
21	292	11039	29
22	2954	18500	14
23	304	9369	14
24	353	7184	28

25	567	10447	16
26	467	5100	13
27	3368	63694	45
28	253	1651	4
28	196	1450	10
30	185	1745	12
31	387	1798	6
32	430	2957	28
33	204	963	6
34	71	1233	6
35	840	3240	6
36	1648	10000	11
37	1035	6800	8
38	548	3850	22
39	2054	14000	31
40	302	5787	26
41	1172	9700	22
42	253	1100	7
43	227	5578	14
44	59	1060	6
45	299	5279	6
46	422	8117	15
47	1058	8710	9
48	65	796	9
19	390	11023	26
50	193	1755	13
51	1526	5931	28
52	575	4456	13
53	509	3600	13
54	583	4557	12
55	315	8752	14
56	138	3440	12
57	257	1981	9

58	423	13700	30
59	495	7105	20
60	622	6816	16
61	204	4620	12
62	616	7451	15
63	3634	39479	33

TABLE 6.8 Software Development Data.

8. Use the R system, found on the Internet and in Appendix G, to run a variety of hierarchical methods on both the original data and the fuzzified data from Problem 7. Compare the solutions between the crisp and fuzzified clustering solutions.

CHAPTER 7

CLASSIFICATION AND ASSOCIATION RULES

In This Chapter

7.1 INTRODUCTION

All of the clustering methods presented so far find the number of clusters and the composition of the clusters which best fit the data. However, many times in an application the clusters already exist and the problem is to assign new data points to the pre-existing clusters, such a procedure is called a classification. Besides, developing classification methods this chapter introduces the concept of association rules. Often classification is accomplished by a tree construction. Given a classification tree enables the construction of If-then, or association, rules.

7.2 DEFINING CLASSIFICATION

What is Classification? Often the groups or clusters are already known and as new objects are posted to the data in order to maintain the organization and categorization of the data, the need is to simply post the data to one of the existing clusters.

To accomplish this task requires maintaining the basic goal of data classification, which is to organize and categorize data into distinct clusters. Model construction based upon the data and associated data distribution must be implemented. Once this task is in place, the model can be used for classifying new data. Basically, prediction is used to decide which existing category in which to place the new data. An example in statistics for only two categories is discriminate analysis. One way to define classification is the following:

Classification = prediction for discrete and nominal values.

Classification falls into two basic types: supervised classification and unsupervised classification. This chapter emphasis is on supervised classification, when the class labels and number of classes are known. In unsupervised classification, the class labels and number of classes are not known, as in the case for clustering.

In general to correctly classify, the following process should be established:

1. Model construction

2. Model evaluation

3. Model application

To develop the model, certain requirements must be supported. First each data pattern or record is to be assumed to belong to an existing cluster through usage of the record's class label value. Most classification models are represented as a set of classification rules, called association rules in the standard "If-then" format or as decision trees.

Model Construction

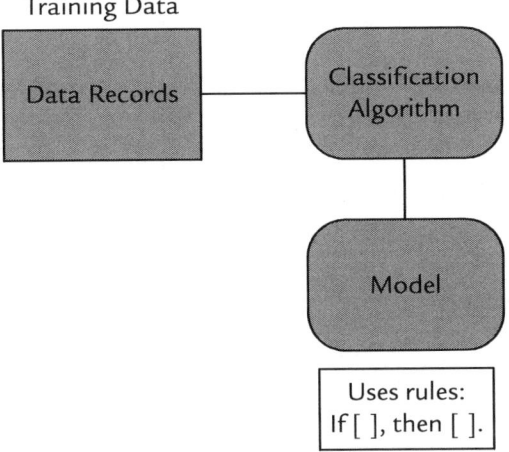

FIGURE 7.1 Model Construction.

During model evaluation, several tests take place. For assessing accuracy one test can be to benchmark the model on a pilot data or test set. Data already in the system can be deleted from their class and then treating this old data as new data, it can be checked to see if the model correctly assigns. The rate of accuracy can be determined by observing the percentage of the test set correctly classified by the model.

Model Construction

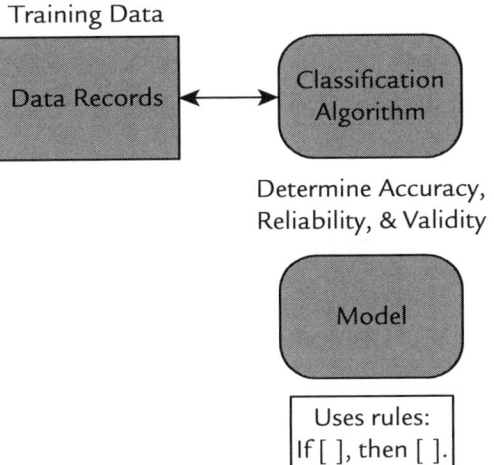

FIGURE 7.2 Functional Interpretation of the Model Evaluation.

Once both model evaluation and model construction have been completed then the model is ready for use. The model should be reliable and valid in:

1. classifying unseen data instances through usage of the class labels, and

2. predicting the actual label values for the data instances.

Many diverse methods of classification have been employed using techniques taken from statistics, artificial intelligence, mathematics, library science, and business among others. A brief list of techniques includes decision tree induction, neural networks, Bayesian classification, association-based classification, K-nearest neighbor, case-based reasoning, genetic algorithms, and fuzzy sets.

7.3 DECISION TREES

Tree structures are ideally suited for classification because rule systems commonly generate sequences of "If [ ], then [ ]" statements. One tree structure often used in artificial intelligence is the decision tree, refer to Figure 7.3. Note that internal nodes represent tests on data attributes and the leaf nodes represent the class labeled either already assigned for old data or to be assigned to the new data. Each branch of the tree represents a test. Branch nodes represent the specific class and all records being posted or searched for possess the same class label value.

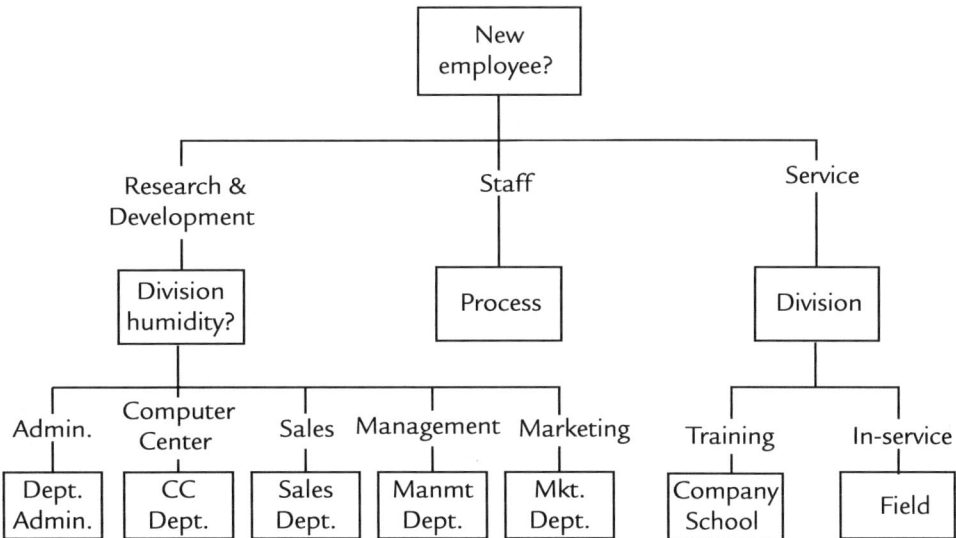

FIGURE 7.3 Decision Tree.

This illustration, as viewed from Figure 7.3, is tied to the concept of decision rules. Each path from the root of the tree to a leaf node represents a sequence of decision rules. Consider the decision tree in Figure 7.4.

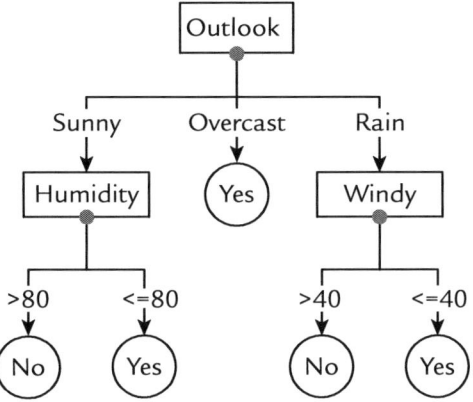

FIGURE 7.4 Relationship Between Decision Rules and Decision Trees.

Figure 7.4 is a famous example from the field of artificial intelligence. It has built into it, by tracing the paths to the leaves, decision rules. The central path is: *If (outlook = overcast) then play (yes)*. The extreme right path is: *If (outlook = sunny) and (humidity <=80) then play (yes)*. The extreme left path is: *If (outlook = rain) and (windy <=40) then play (yes)*. There are two more rules.

1. Classifying new data: look at the data record's attribute value for the feature specified. Move along the edge labeled with this value.

2. If you reach a leaf, return the label of the leaf.

3. Otherwise, repeat from step 1.

FIGURE 7.5 Decision Tree Algorithm for Classification.

7.4 ID3 TREE CONSTRUCTION ALGORITHM

Tree construction at the start involves partitioning of the test data with known attribute values. The rules are implemented from the root down. The taller the trees and the maximum number of edges from the root to a leaf node, the less efficient storage and searching will become. Additionally, many seldom or never used rules will be represented by some path in the tree. Again the storage, posting of new data, and searching times may be hampered. The solution to these problems is to perform tree pruning at tree construction time. Basically, tree pruning involves removing tree branches that may reflect noise in the training data and lead to errors when classifying test data.

The basic steps in tree construction are:

1. The tree starts as a single node representing all data.

2. If sample data are all in the same class, then the node becomes a leaf labeled with the class label.

3. Otherwise, select the feature that best separates sample data into individual classes.

4. Recursion stops when:

    a. Samples in the node belong to the same class (majority), or

    b. There are no remaining attributes on which to split.

FIGURE 7.6 Basic ID3 Trees Construction Algorithm.

Input: a set of examples S, a set of features F, and a target set T (target class T represents the type of instance we want to classify, e.g., whether "to play golf")

1. If every element of **S** is already in **T**, return "yes"; if no element of **S** is in **T** return "no".

2. Otherwise, choose the best feature *f* from **F** (if there are no features remaining, then return failure).

3. Extend tree from *f* by adding a new branch for each attribute value.

4. Distribute training examples to leaf nodes (so each leaf node **S** is now the set of examples at that node, and **F** is the remaining set of features not yet selected).

5. Repeat Steps 1–5 for each leaf node.

Main Question:

How do we choose the best feature at each step?

Note: ID3 algorithm only deals with categorical attributes, but can be extended.

The ID3 algorithm, described in Whitten and Frank,[1] constructs a decision tree by taking a set of training data and arranging the feature and feature values into a tree structure as described previously. ID3 addresses the determination of the order of the features examined, the organization of the decision tree, by using an entropy-based metric called Information Gain (IG) that selects the attribute which will best separate the instances into subsets representing a single class. A series of improvements to ID3 have culminated in an influential and widely used system for decision tree induction called C4.5. C4.5 appears in a classic book by Quinlan,[2] which gives a listing of the complete C4.5 system, written in the C programming language. The latter improvements include methods for dealing with numeric attributes, missing values, noisy data, and generating rules from trees.

Currently, two main methods to construct fuzzy decision trees are popular. The first method is based upon the application of a generalized

---

[1] Whitten, I. H., & Frank, E. (2000). Divide and Conquer: Constructing Decision Trees. *Data Mining: Practical Machine Learning Techniques with Java Implementations*. San Francisco, CA: Morgan Kaufmann Publishers, pp. 49-50.

[2] Quilan, J. R. (1986). Induction of Decision Trees. *Machine Learning*, Vol. 1, pp. 81-106. (1990). Decision Trees and Decision Making. *IEEE Transactions on Systems, Man, and Cybernetics*, Vol. 20, no. 2, pp. 339-346.

Shannon entropy, or the entropy of future events as a measure of discrimination (Bouchon-Meunier, Marsala, and Ramdani),[3] (Janikow),[4] and (Weber).[5] This method substitutes probabilities of fuzzy events to classical probabilities. The second method is based upon another family of fuzzy measures (Cios and Sztanadera)[6] and (Wang, Qian, and Ye).[7]

7.4.1 Choosing the "Best" Feature

Entropy, $E(I)$ of a set of instance I, containing p positive and n negative examples

$$E(I) = -\frac{p}{p+n} log_2 \frac{p}{p+n} - \frac{n}{p+n} log_2 \frac{n}{n+p}.$$

Gain(A, I) is the expected reduction in entropy due to feature A

$$\text{Gain}(A, I) = E(I) - \sum_{descendant} \frac{p_j + n_j}{p+n} E(I_j)$$

the jth descendant of I is the set of instances with value v_j for A.

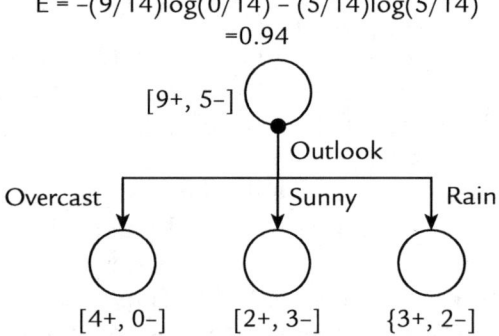

E = –(9/14)log(0/14) – (5/14)log(5/14)
=0.94

FIGURE 7.7 Finding the "Best" (Most Discriminating) Feature.

[3] Bouchon-Meunier, B., Marsala, C., & Ramdani, M. (1997). Learning from Imperfect Data. *Fuzzy Information Engineering: A Guided Tour of Applications*, D. Dubois, H. Prade, and R. R. Yager, editors, John Wiley and Sons, pp.139-148.

[4] Janikow, C. (1998). Fuzzy Decision Trees: Issues and Methods. *IEEE Transactions on Systems, Man, and Cybernetics*, Vol. 28, no. 1, pp. 1-14.

[5] Weber, R. (1992). Fuzzy-ID3: A Class of Methods for Automatic Knowledge Acquisition. *IIZUKA'92 Proceedings of the 2nd International Conference on Fuzzy Logic*, pp. 265-268.

[6] Cios, K., & Sztandera, L. (1992). Continuous ID3 Algorithm with Fuzzy Entropy Measures. *Proc. Of The First International IEEE Conference on Fuzzy Systems*.

[7] Wang, X., Chen, B., Qian, G., & Ye, F. (2000). On the Optimization of Fuzzy Decision Trees. *Fuzzy Sets and Systems*, Vol. 112, no. 1, pp. 117-125.

7.4.2 Information Gain Algorithm

Assume a universe of objects described by N training instances. This universe of objects is completely described by n attributes and one class attribute. This set of attributes is defined as $A = \{A^{(1)}, A^{(2)}, \ldots, A^{(n)}\}$. For the kth attribute, $A^{(k)}$, there are m_k possible attribute values that define the subsets which can be classified into C subsets $\omega_1, \omega_2, \ldots, \omega_C$.

The metric used to determine the splitting attribute is based on the classical definition of entropy. $H_i^{(k)}$ is the entropy extant in the ith subset of a set split on attribute $A^{(k)}$ and is defined as

$$H_i^{(k)} = \Sigma_{j=1} C - p_i^{(k)}(j) \log(p_i^{(k)}(j)),$$

where $p_i^{(k)}(j)$ is the relative frequency of the ith subset of the kth attribute with respect to the subset ω_j where $1 \leq j \leq C$. That is, $H_i^{(k)}$, represents the entropy present in the child node created by those instances that have the ith value of the proposed splitting attribute $A^{(k)}$. The relative frequency, in turn, is defined as

$$p_i^{(k)}(j) = \{M(A_i^{(k)} \cap \omega_j)\}/\{M(A_i^{(k)})\},$$

where $A_i^{(k)}$ represents all instances that have the ith value of attribute $A^{(k)}$, $A_i^{(k)} \cap \omega_j$ representing the members of ω_j that have the ith value of attribute $A^{(k)}$ and $M(.)$ is the cardinality of a fuzzy set. Simply stated, $p_i^{(k)}(j)$ is the proportion of instances in the child node that are members of ω_j.

In order to choose the best splitting attribute, the sum of the entropy for all nodes that would be created by splitting on that attribute must be computed. $E^{(k)}$ is the combined entropy extant in all subsets created by splitting on $A^{(k)}$ and is defined as

$$E^{(k)} = \Sigma_{i=1} {}_k^m \{[M(A_i^{(k)})/\Sigma_{j=1} {}_k^m M(A_j^{(k)})] * H_i^{(k)}\},$$

where $\Sigma_{j=1} {}_k^m M(A_j^{(k)})$ is the cardinality of all instances in the node being split. The combined entropy $E^{(k)}$, therefore, computes the sum of the entropy present at each child node weighted by the proportion of instances that occupy the child node.

An attribute, $A^{(h)}$, is chosen for splitting at each non-leaf node such that h is the value of the index k that corresponds to the minimum entropy as a result of splitting on $A^{(k)}$. That is $A^{(h)}$ is the attribute that corresponds to $\min(E^{(k)})$ for $1 \leq k \leq n$.

In practice, no further calculations are needed. The attribute that minimizes the combined entropy of the child nodes is the attribute that produces the best localized split, forming the most unambiguously classified nodes even if no further splitting were to take place. The Information Gain metric, however, derives its name from the difference of entropy extant in the current node, E, and the combined entropy present in its child nodes, $E^{(k)}$. The entropy of the current node is defined as

$$E = \Sigma_{j=1}^{C} - p(j) \log(p(j)),$$

where $p(j) = M(\omega_j)/M(N)$ represents the frequency of an instance in the node belonging to the class ω_j. Using this definition, the Information Gain, or $IG^{(k)}$, for splitting on the attribute $A^{(k)}$ can be calculated as

$$IG^{(k)} = E - E^{(k)}.$$

This illustrates that $IG^{(k)}$ is the expected reduction in entropy caused by knowing the value of $A^{(k)}$. The attribute that maximizes the decrease in entropy, or maximizes the gain in information, therefore, minimizes the amount of information needed to classify the instances in the resulting child nodes (Doug and Kothari).[8]

Clearly, using Information Gain as an attribute selection measure minimizes the expected levels of branching needed to classify an instance, although this method does not ensure that the simplest tree will be formed.

Several data sets are available for testing the classification accuracy for the fuzzy ID3 tree induction algorithm. Nominal and numeric and data sets containing both nominal and numeric attributes can be selected from the UCI Machine Learning Repository (Blake and Metz).[9] The example of the weather problem is from Quinlan[2] and has been widely used to explain machine learning schemes. This repository is a publicly available collection of databases submitted from data sources worldwide. The Iris data set, which dates back to seminal work by the eminent statistician R. A. Fisher in the mid-1930s and is arguably the most famous data set used in data mining (Fisher).[10] These data sets can be obtained from a source online.

---

[8] Doug, M., & Kothari, R. (2001). Look-Ahead Based Fuzzy Decision Tree Induction. *IEEE Transactions on Fuzzy Systems*, 9(3), 461-468.

[9] Blake, C. L., & Merz, C. J. (1998). UCI Repository of Machine Learning Databases. Department of Information and Computer Sciences, University of California, Irvine, CA URL: http://www.ics.uci.edu/~mlearn/MLRepository.html.

[10] Fisher, R. (1936). The use of Multiple Measurements in Taxonomic Problems. *Annual Eugenics 7 (part II)*, reprinted in *Contributions to Mathematical Statistics*, 1950, pp. 179-188.

Day	Outlook	Temp	Humidity	Wind	Play
D1	sunny	hot	high	weak	No
D2	sunny	hot	high	strong	No
D3	overcast	hot	high	weak	Yes
D4	rain	mild	high	weak	Yes
D5	rain	cool	normal	weak	Yes
D6	rain	cool	normal	strong	No
D7	overcast	cool	normal	strong	Yes
D8	sunny	mild	high	weak	No
D9	sunny	cool	normal	weak	Yes
D10	rain	mild	normal	weak	Yes
D11	sunny	mild	normal	strong	Yes
D12	overcast	mild	high	strong	Yes
D13	overcast	hot	normal	weak	Yes
D14	rain	mild	high	strong	No

TABLE 7.1 The Weather Data Set.

By applying the formulas in Figure 7.8 to the weather data set, we can find the gain in information for branching from the humidity to the next level of the tree.

$$E(I) = -\frac{p}{p+n} log_2 \frac{p}{p+n} - \frac{n}{p+n} log_2 \frac{n}{n+p}$$

and

$$Gain(A, I) = E(I) - \sum_{descendant} \frac{p_j + n_j}{p+n} E(I_j)$$

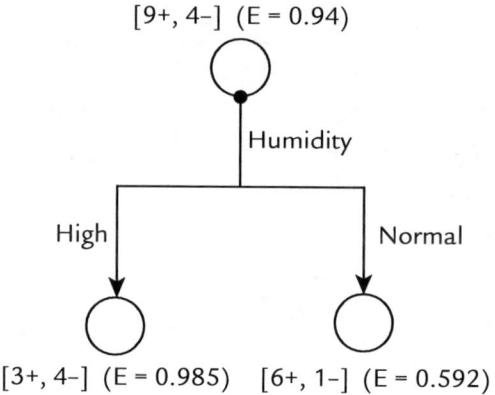

[9+, 4–] (E = 0.94)

Humidity

High   Normal

[3+, 4–] (E = 0.985)   [6+, 1–] (E = 0.592)

Gain (humidity) = 0.94 – (7/14)*0.985 – (7/14)
*0.592) = 0.151

FIGURE 7.8 Gain in Information for Humidity Node Branching.

For the wind node, the gain in information is:

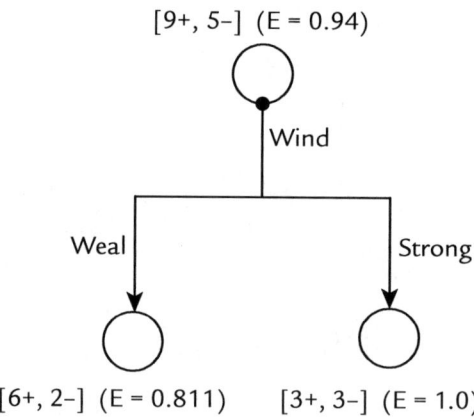

[9+, 5–] (E = 0.94)

Wind

Weal   Strong

[6+, 2–] (E = 0.811)   [3+, 3–] (E = 1.0)

Gain(humidity) = 0.94 – (8/14)*0.811 – (8/14)
*1.0) = 0.048

FIGURE 7.9 Gain in Information for Wind Node Branching.

Because outlook has the largest gain in information, then outlook would be selected as the root in the path to a leaf node.

Sometimes during the construction of a classification tree, the tree overfits the training data set. Overfitting is due to noise being present in the data set. Another explanation for overfitting is that the training data was too small and, as a result, some of the actual rules were not present in the training data set. One solution to overfitting is of a preventive nature, simply stop growing the tree at an earlier time. A post hoc method for avoiding overfitting is to postpone the corrective action until tree pruning is performed.

Another question that must be answered is Has the "correct" tree been found? One method for finding the "correct" tree is to use all the data for training and at each branch node during tree construction determine whether splitting nodes in the current tree structure would enhance the overall information gain. Many times the decision to split nodes can be made during pruning.

7.4.3 Tree Pruning

Pruning is the process of removing branches and subtrees that are generated due to noise, which improves classification accuracy. The final version of the ID3 tree construction, a recursive process utilized at each node starting with the root node, is described below:

1. **Create Node.** Create a new node that represents all instances that have made it to this point.

2. **Test Stopping Criteria.** Test the stopping criteria to determine whether this node will be a leaf node or an internal node that requires further splitting.

Stopping Criteria

- If all instances belong to class ω_j, then return the current node as a leaf node labeled with the class ω_j.

- If the attribute list is empty, compute the class distribution and return the current node as a leaf node labeled with the most prevalent class by the majority voting.

- If no instances are present at this node, then return a null node, a leaf node that is unclassified.

Determine Splitting Attribute. Compute the Information Gain of each attribute, $A^{(k)}$, in the list of attributes, and assign the attribute for this node.

i. Label Current Node. Label the current node with the splitting attribute.

Partition Data. For each known value i, $1 \leq i \leq m_h$, retrieve only the instances that have the ith value of $A^{(h)}$ from the list of attributes for each of the data partitions because this attribute has now been used as a splitting criterion and no further information can be gained by using it again.

Create Child Nodes. For each subset of instances created by splitting the data, create a child node by making a recursive call to (2). Label the pathway to each new node with the appropriate value for $A^{(h)}$.

FIGURE 7.10 Final ID3 Tree Construction Algorithm.

7.5 BAYESIAN CLASSIFICATION

Bayesian classification is another method of classifying using probability instead of decision tree construction. This method is a statistical classifier based on Bayes theorem. It uses probabilistic learning by calculating explicit probabilities for the hypothesis. By assuming total independence between attributes, a Bayesian classifier can be used with a high degree of confidence with large data sets. Each training example can incrementally increase or decrease the probability that a hypothesis is correct. The assumption that prior knowledge can be combined with observed data is enforced.

Example:

Given a standard deck of fifty-two playing cards. Let B1 be the event of drawing the ace of spades. Let B2 be the event of drawing a face card in spades. Find the probability of B1 given that B2 has already happened.

- The question above wants $P(B1|B2)$.

- The answer intuitively wants the probability of drawing the ace of spades only if the sample space after B2 actually takes place. This revises the sample space to contain only 9 playing cards.

- So the answer is 1/9, not 1/13 or the $P(B1)$.

- The equation to solve it is:

$$P(B1|B2) = P(B1{\cap}B2)/P(B2) \text{ [Product Rule]}$$

$$P(B1 \text{ and } B2) = P(B1) * P(B2) \text{ [If B1 and B2 are independent]}$$

In general, $P(A{\cap}B) = P(B) \times P(A|B)$ and $P(B{\cap}A) = P(A) \times P(B|A)$ for any two events A and B. Since $P(A{\cap}B) = P(B{\cap}A)$, then $P(B) \times P(A|B) = P(A) \times P(B|A)$. Therefore the following holds: $P(A|B) = [P(A) \times P(B|A)] / P(B)$, which implies the result:

$$P(A|B) = \frac{P(A)P(B|A)}{P(B)} = \frac{P(A)P(B|A)}{P(A)P(B|A) + P(\bar{A})P(B|\bar{A})}$$

The general Bayesian theorem is:

Given E1, E2,…,En are mutually **disjoint** events and $P(Ei) \neq 0$, ($i = 1, 2,…, n$)

$$P(Ei/A) = [P(Ei) \times P(A|Ei)] / \Sigma [P(Ei) \times P(A|Ei)].$$

Consider being given a data sample X with an unknown class label where H is the hypothesis that X belongs to a specific class C. The *conditional probability* of hypothesis H given X, $Pr(H|X)$, follows Bayes theorem:

$$Pr(H/X) = \frac{Pr(X/H)\,Pr(H)}{Pr(X)}.$$

Suppose we have n classes $C_1, C_2, \ldots, C_n$. Given an unknown sample X, the classifier will predict that $X = (x_1, x_2, \ldots, x_n)$ belongs to the class with the highest conditional probability.

$$X\varepsilon C_i \text{ if } Pr\left(C_i/X\right) > Pr\left(C_j/X\right), \text{ for } i \le j \le n, i \ne j$$

Note that maximizing $Pr(X/C_i) * Pr(C_i)/Pr(X)$ is in reality maximizing $Pr(X/C_i)\,Pr(C_i)$. Then the computations for the Bayesian classifier are:

$$Pr(C_i) = s_i/s \text{ and}$$

$$Pr(X/C_i) = \prod_{k=1}^{n} Pr(x_k/C_i) \text{ where } Pr(x_k/C_i) = s_k/s_i.$$

Why use Bayesian Classification?

- **Probabilistic learning:** Calculate explicit probabilities for hypothesis, among the most practical approaches to certain types of learning problems.

- **Incremental:** Each training example can incrementally increase/decrease the probability that a hypothesis is correct. Prior knowledge can be combined with observed data.

- **Probabilistic prediction:** Predict multiple hypotheses, weighted by their probabilities.

- **Standard:** Even when Bayesian methods are computationally intractable, they can provide a standard of optimal decision making against which other methods can be measured.

A simplified assumption when constructing a Naïve Bayesian Classifier (NBC) is that the attributes are conditionally independent. This assumption reduces the computation cost, because the only count required is the class distribution. The probabilistic model of NBC is to find the probability of a certain class given multiple disjoint (assumed) events.

The NBC applies to learning tasks where each instance x is described by a conjunction of attribute values and where the target function $f(x)$ can take on any value from some finite set V. A set of training examples of the target function is provided, and a new instance is presented, described by the tuple of attribute values <a1, a2,..., an>. The learner is asked to predict the target value, or classification, for this new instance.

Abstractly, probability model for a classifier is a conditional model: $P(C|F_1, F_2, ..., F_n)$. Over a dependent class variable C with a small number of outcome or classes conditional over several feature variables $F_1, ..., F_n$.

The Naïve Bayes Formula is:

$$P(C|F_1, F_2,..., F_n) = \text{argmax}_c [P(C) \times P(F_1|C) \times P(F_2|C) \times ... \times P(F_n|C)] / P(F_1, F_2,..., F_n)$$

Because $P(F_1, F_2, ..., F_n)$ is common to all probabilities, we do not need to evaluate the denominator for comparisons.

The following discussion builds an NBC for the following data set:

Age	Income	Student	Credit_rating	Buys_computer
$x \leq 30$	High	No	Fair	No
$x \leq 30$	High	No	Excellent	No
$30 < x \leq 40$	High	No	Fair	Yes
$x > 40$	Low	Yes	Fair	Yes
$x > 40$	Low	Yes	Excellent	No
$30 < x \leq 40$	Low	Yes	Excellent	Yes
$x \leq 30$	Medium	No	Fair	No
$x \leq 30$	Low	Yes	Fair	Yes
$x > 40$	Medium	Yes	Fair	Yes
$x \leq 30$	Medium	Yes	Excellent	Yes
$30 < x \leq 40$	Medium	Yes	Excellent	Yes
$30 < x \leq 40$	High	Yes	Fair	Yes
$x > 40$	Medium	No	Excellent	No

TABLE 7.2 Sample Data Set.

We are given a new data point, x = (age$_i$, income$_i$, student$_i$, credit-rating$_i$) and need to estimate $P(buy\_computer|x)$ in order to post the new data point into the data set.

Note that $P(buy\_computer|x_i) = \dfrac{P(x_i|buy\_computer)P(buy\_computer)}{P(x_i)}$.

Therefore we need to choose the value of buy_computer that maximizes $P(buy\_computer|x_i)$. This means we need to choose the value of buy_computer that maximizes $P(x_i|buy\_computer)P(buy\_computer)$.

Assume for $(x_1, x_2, ..., x_4)$ *to be posted in X after determination of the value for buy_computer that*

$$P(x_i|buy\_computer_j) = P(x_1|buy\_computer_j) * P(x_2|buy\_computer_j) * ... * P(x_4|buy\_computer_j)$$

The first step is to compute $P(x_i|buy\_computer_j)$ for all x_i and $buy\_computer_j$.

The new data point is classified as a specific $buy\_computer_j$ value if

$P(buy\_computer_j)\prod_j P(x_i|buy\_computer_j)$ is maximal.

The classes for buy_computer has the following probabilities: $P(buy\_computer = No) = \dfrac{5}{13}$ and $P(buy\_computer = Yes) = \dfrac{8}{13}$.

$P(x_1 = x \leq 30|buy\_computer = No) = 3/5$; $P(x_1 = x \leq 30|buy\_computer = Yes) = 2/8$

$P(30 < x \leq 40|buy\_computer = No) = 0$; $P(30 < x \leq 40|buy\_computer = Yes) = 2/8$

$P(x > 40|buy\_computer = No) = 2/5$; $P(x > 40|buy-computer = No) = 4/8$

$P(Student = No|buy\_computer = No) = 4/5$;

$P(Student = No|buy\_computer = Yes) = 1/8$;

$P(Student = Yes|buy\_computer = No) = 1/5$;

$P(Student = Yes|buy\_computer = Yes) = 7/8$;

$P(Income = High|buy\_computer = No) = 2/5$;

$P(Income = High|buy\_computer = Yes) = 2/8$;

$P(Income = Medium|buy\_computer = No) = 1/5$;

$P(Income = Medium|buy\_computer = Yes) = 4/8$;

$P(Income = Low|buy\_computer = No) = 1/5$;

$P(Income = Low|buy\_computer = Yes) = 3/8$;

P(*Credit-rating* = Fair|*buy_computer* = No) = 2/5;

P(*Credit-rating* = Fair|*buy_computer* = Yes) = 5/8;

P(*Credit-rating* = Excellent|*buy_computer* = No) = 3/5;

P(*Credit-rating* = Excellent|*buy_computer* = Yes) = 3/8;

Then P(x|*buy_computer* = No) = P(Age is $x \le 30$|*buy_computer* = No) *

P(*Income* = Medium|*buy_computer* = No) *

P(*Student* = No|*buy_computer* = No) *

P(*Credit-rating* = Fair|*buy_computer* = No)

P(x|*buy_computer* = No) = 0*(1/5)(4/5)(2/5)

Note that if one of the conditional probabilities is zero, then the entire expression is zero. To overcome this situation, use the estimate

$$P(x_i | buy\_computer = No) = \frac{N_{i,buy\text{-}computer=No} + 1}{N_{buy\text{-}computer=No} + Number\ of\ classes} = 3/7$$

Then P(x|*buy_computer* = No) = (3/7)(1/5)(4/5)(2/5) = 0.0274

P(x|*buy_computer* = Yes) = P(Age is x = 30|*buy_computer* = Yes) *

P(*Income* = Medium|*buy_computer* = Yes) *

P(*Student* = No|*buy_computer* = Yes) *

P(*Credit_rating* = Fair|*buy_computer* = Yes)

P(x|*buy_computer* = Yes) = (1/8)(3/8)(1/8)(4/8) = 0.0293

P(x|*buy_computer* = No) P(*buy_computer* = No) = (0.0274)(5/13) = 0.0105

P(x|*buy_computer* = Yes) P(*buy_computer* = Yes) = (0.0293)(8/13) = 0.018

Then P(*buy_computer* = No|x) $\le$ P(*buy_computer* = Yes|x) implies that x is assigned to the class *buy_computer* = Yes.

Note that in this problem instance all the attributes were categorical. For each continuous attribute A in a data set for an NBC, simply implement a set of labels for the continuous values. Place one ordinal value per label. By assuming the data is normally distributed, which allows the use of the sample mean and standard deviation as estimates for the population mean and standard deviation, the P(the jth attribute value |C) can be found.

7.6 ASSOCIATION RULES

We have seen how a classification tree generates If-Then rules, which are referred to as association rules. Actually association rules were developed for conducting a market basket analysis. Grocery stores need to find answers to the following questions: What do the customers purchase? Are certain combinations of products purchased in a given transaction? Basically, management needs to determine the associations and correlations between the items commonly found in their shopping baskets.

A *transaction database*, T, is a finite set of transactions $T = \{t_1, t_2, \ldots, t_n\}$, where each transaction contains a set of items, I. An *item set* is a finite set of items, $I = \{s_1, s_2, \ldots, s_m\}$. The goal is to determine frequent patterns, associations, correlations, or causal structures contained in the item sets in the transaction database and express these relationships in terms of *association rules*, If-Then rules.

Consider the following transactions for purchases at the local grocery store, as illustrated in Table 7.3:

Trans. ID	Items
T1	Soda, potato chips, chip dip
T2	Soda, bread, meat
T3	Bread, mustard, meat
T4	Soda, cookies
T5	Bread, potato chips
T6	Bread, meat, ketchup

TABLE 7.3 Grocery Store Transaction Database.

The frequency support count for an item set, $\sigma(I)$, is the frequency of occurrence of the item set across the transaction database. For the latter transactions, the following support counts are present:

$$\sigma(\{\text{soda, potato chips}\}) = 1$$

$$\sigma(\{\text{soda}\}) = 3$$

The support for an item set, s(I), is the fraction of transactions in the transaction database that contain the item set. For the transactions in Table 7.3 see the following:

$$s(\{soda, potato\ chips\}) = 1/6$$

$$s(\{soda\}) = 3/6$$

A *frequent item set* is an item set whose support is greater than or equal to a minimum support threshold, the minsup. For example, if we set the minsup to 1/3 then {soda} is a frequent item set and {soda, potato chips} is not a frequent item set.

An *association rule* is an implication between two item sets. For example, in the transaction database in Table 7.3 we note that, if the customer purchases meat, then the customer purchases meat in the same transaction. The association rule is denoted as:

{meat} → {bread}.

Note that the occurrence of the association rule is:

$$S(X \cup Y) = \frac{\sigma(X \cup Y)}{k},$$ where k is the number of transactions that contain both X and Y.

For {meat} → {bread}:

$$s(\{meat\} \rightarrow \{bread\}) = 3/4 = 0.75$$

The confidence of the association rule $X \rightarrow Y$, or the strength of the association, is a measurement of how often items in the item set Y appear in transactions that contain the item set X.

$$C(X \rightarrow Y) = \frac{\sigma(X \cup Y)}{\sigma(X)}.$$

For {meat} → {bread}:

$$C(For\ \{meat\} \rightarrow \{bread\}) = 4/3 = 1.33.$$

Many distinct types of rules exists:

1. Binary association rules:

{meat} → {bread}

2. Quantitative association rules:

US currency [\$25] → Old English currency [10 pounds].

3. Fuzzy association rules:

Tall → Heavy

To thoroughly study the transaction database in Table 7.3 we need to consider all combinations of items. Because there are 8 items we would need to consider

$$\binom{8}{0} + \binom{8}{1} + \ldots + \binom{8}{8} = 256 \text{ item sets.}$$

These item sets generate 512 association rules to consider. Do not panic! There exists a downward closure property to enable us to only have to investigate a proper subset of the total set of items.

Downward Item Set Closure Theorem: *Any subsets of a frequent item set are also frequent item sets.*

This leads to the following algorithm to determine all the frequent item sets in a transaction database, called the *Apriori Algorithm:*

Apriori Algorithm

Step 1: $k = 1$

Step 2: Generate all frequent item sets of length 1.

Step 3: Repeat until no frequent item sets are found.

    $k: = k + 1$

    Generate item sets of size k from the $k - 1$ frequent item sets

    Calculate the support of each candidate by scanning the transaction database.

FIGURE 7.11 The Apriori Algorithm.

In summary, the search for all association rules consists of two primary steps:

1. Generate all frequent item sets whose support is greater or equal to minsup.

2. Use the frequent item sets to generate association.

For each frequent item set I, we can search for all nonempty subsets A of I, such that the association rule I $\rightarrow$ I – A satisfies the minimum confidence. Then I $\rightarrow$ I – A is a rule.

If I = {A, B, C} then the trivial association rules hold: AB $\rightarrow$ C, AC $\rightarrow$ B, BC $\rightarrow$ A, A $\rightarrow$ BC, B $\rightarrow$ AC, and C $\rightarrow$ AB. There are $2^k - 2$ candidate association rules, where k is the number of items in the item set I. Also note that for a given item set I = {A, B, C, D} then if c(ABC $\rightarrow$ D) $\geq$ c(AB $\rightarrow$ CD) $\geq$ c(A $\rightarrow$ BCD). This last property for item sets can be used for pruning during rule generation.

7.7 PRUNING

Why prune? The answer is due to overfitting. This is due to the construction of a tree with too many branches, some of which may reflect anomalies due to noise or outliers. The overfitting could be due to a nonrepresentative input data set during tree construction.

There are two approaches to tree pruning:

1. Prepruning: Halt tree construction early if splitting a node would result in the information gain falling below a threshold value.

2. Postpruning: Remove branches after the completion of tree construction and then prune the tree.

The ID3 algorithm can grow each branch of the tree just deeply enough to perfectly classify the examples based upon the input data set. This practice can lead to difficulties when there is noise in the data, or when the data set is too small to produce a representative sample of the true target function. In both cases, the ID3 algorithm can produce trees that *overfits*.

There are several approaches to avoiding overfitting in decision tree learning. These can be grouped into two classes:

- approaches that stop growing the tree earlier, before it reaches the point where it perfectly classifies the data,

- approaches that allow the tree to overfit the data, and then post prune the tree.

Although the first of these approaches might seem more direct, the second approach of postpruning overfit trees has been found to be more successful in practice. This is due to the difficulty in the first approach of estimating precisely when to stop growing the tree. Regardless of whether the correct tree size is found by stopping early or by postpruning, a key question is what criterion is to be used to determine the correct final tree size. Approaches include:

- Use a separate data set, distinct from the data set used in tree construction, to evaluate the utility of postpruning nodes from the tree.

- Use all the available data for tree construction, but apply a statistical test to estimate whether expanding (or pruning) a particular node is likely to produce an improvement beyond the rules extracted from the tree construction data set.

- Use an explicit measure of the complexity for encoding the association rules constructing the tree, halting growth of the tree when this encoding size is minimized. This approach is based on a heuristic called the Minimum Description Length principle.

The first of the above approaches is the most common and is often referred to as a training and validation set approach. In this approach, the available data are separated into two sets of examples: a *rule construction set*, which is used to form the learned hypothesis, and a separate *validation set*, which is used to evaluate the accuracy of this hypothesis over subsequent data and, in particular, to evaluate the impact of pruning this hypothesis.

7.8 EXTRACTION OF ASSOCIATION RULES

The extraction rules are:

1. Represent the knowledge in the form of If-Then rules.

2. Each rule is created for each path from the root to a leaf.

3. Each attribute-value pair along a path forms a conjunction.

7.9 SUMMARY

- The ID3 tree induction algorithm is well-suited for analyzing and condensing purely nominal data sets into an easily interpreted tree structure that can be used for making rules and classifications of unknown instances.

▪ A fuzzy modification of the ID3 tree which combines the simplicity of data representation in a tree structure would enable the ability to analyze data sets that contain both nominal and numeric attributes.

▪ Fuzzy decision trees would allow for the maintaining the ease of interpretation of a decision tree while increasing the flexibility of its representation, so that analytic tools can be employed in their construction. By including fuzzy tests in the decision tree construction, it would increase its expressive capability and incorporate into its construction characteristics of connectionist methods that are useful to generate a learning device that is flexible, robust, and capable to generalize.

▪ A statistical classification system:

  1. Defines classes of objects

  2. Specifies probability distribution model connecting classes to observable features

  3. Fits parameters of model to data

  4. Observes features on inputs and compute probability of class membership

  5. Assigns object to a class

▪ A Bayesian classification system:

  1. Defines classes

  2. a. Specifies probability model

     b. And prior distribution over parameters

  3. Finds posterior distribution of model parameters, given data

  4. Computes class probabilities using posterior distribution (or element of it)

  5. Classifies objects

▪ The advantages of a Bayesian classifier include:

  • Simple to implement

    – No numerical optimization, matrix algebra, etc.

- Efficient to train and use
  - Fitting is accomplished by computing means of feature values
  - Easy to update with new data
  - Equivalent to linear classifier, so fast to apply
- Independence allows parameters to be estimated on different data sets, e.g.,
  - Estimate content features from messages with headers omitted
  - Estimate header features from messages with content missing
- Generative model
  - Comparatively good effectiveness with small training sets
  - Unlabeled data can be used in parameter estimation (in theory)
- The disadvantages for Bayesian classifiers include:
  - Independence assumption wrong
    - Absurd estimates of class probabilities
    - Threshold must be tuned, not set analytically
  - Generative model
    - Generally lower effectiveness than discriminative techniques (e.g., log, regress)
    - Improving parameter estimates can *hurt* classification effectiveness

7.10 EXERCISES

Use the UCI Machine Learning Repository (Blake and Metz),[9] found on the Internet, as the data set for a particular problem. This repository is a publicly available collection of databases submitted from data sources worldwide. In particular obtain access to the following data sets: Nominal Weather Data, Mixed Weather Data, Nominal Car Data, Numeric Iris Data, and Numeric Yeast Data.

The first Nominal Weather Data is based upon a small, nominal data set. This data set is used primarily for illustrative purposes, to highlight applying the ID3 tree induction algorithm for purely nonnumeric instances where the fuzzification of features is not required.

The Mixed Weather Data consists of fourteen instances of two nominal attributes, outlook and wind, and two numeric attributes, temperature and humidity, plus a classification attribute, play. A determination is made as whether or not the environmental conditions are suitable for playing outside.

Seven nominal attributes, in the nominal car data set, describing aspects of a car's maintenance, physical features, safety, and cost are used to determine the degree to which a car would be considered a "good" buy. There are four possible classifications: unacceptable, acceptable, good, and very good. In total, the data set consists of 1,728 instances.

The well-known iris data set involves only numeric instances. In addition to having four numeric attributes describing the length and width of both the sepal and petal of three species of irises, the iris data set is also known to be nonlinearly separable. A determination of the correct species (Iris-setosa, Iris-veriscolor, or Iris-virginica) of the iris given the four measurements is made.

The numeric yeast data set originates from the Osaka University Institute of Molecular and Cellular Biology. Each instance represents a protein sample, and the goal of the classification is to determine from which cellular location the protein sample originated inside a yeast cell. The nine numeric attributes quantify biochemical tests performed on the protein samples. Each of the 1,484 instances can be classified into one of ten possible cellular locations.

1. Manually construct an ID3 classification tree for the nominal weather data set.

2. Compute the information gain in the humidity node of the ID3 classification tree for the weather data set.

3. Express the decision tree for Problems 1 and 2 as a set of rules.

4. Either develop and implement ID3 classification tree software or use software you have access to for constructing the ID3 classification tree for the following data set.

Age	Income	Student	Credit_rating	Buys_computer
$x \leq 30$	High	No	Fair	No
$x \leq 30$	High	No	Excellent	No
$30 < x \leq 40$	High	No	Fair	Yes
$x > 40$	Low	Yes	Fair	Yes
$x > 40$	Low	Yes	Excellent	No
$30 < x \leq 40$	Low	Yes	Excellent	Yes
$x \leq 30$	Medium	No	Fair	No
$x \leq 30$	Low	Yes	Fair	Yes
$x > 40$	Medium	Yes	Fair	Yes
$x \leq 30$	Medium	Yes	Excellent	Yes
$30 < x \leq 40$	Medium	Yes	Excellent	Yes
$30 < x \leq 40$	High	Yes	Fair	Yes
$x > 40$	Medium	No	Excellent	No

TABLE 7.4 Hypothetical Sample Data.

5. Redo Problem 3 for the nominal yeast data set. Note this data set contains only numeric attributes. To work with continuous variables, use the following strategy:

   1. sort the examples according to the continuous attribute A

   2. identify adjacent examples that differ in their target classification

   3. generate a set of candidate thresholds midway

   problem: may generate too many intervals

   or

   1. take a minimum threshold M of the examples of the majority class in each adjacent partition; then merge adjacent partitions with the same majority class

6. Build a Bayesian classifiers for the weather data sets.

7. Build a Bayesian classifier for the following data file:

Id	Age	Sex	Region	Income	Married	Children	Car	Save_act	Current_act	Mortgaage	Pep
ID12101	48	female	inner_city	17546	no	1	no	no	no	no	yes
ID12102	40	male	Town	30085.1	yes	3	no	no	yes	yes	no
ID12103	51	female	inner_city	16575.4	yes	0	yes	yes	yes	no	no
ID12104	23	female	Town	50576.3	yes	3	no	no	yes	no	no
ID12105	57	female	Rural	20375.4	yes	0	yes	yes	no	no	no
ID12106	57	female	Town	67869.6	yes	2	yes	yes	yes	no	no
ID12107	22	male	Rural	8877.07	no	0	no	no	yes	no	yes
ID12108	58	male	Town	24946.6	yes	0	yes	yes	yes	no	no
ID12108	37	female	SUBURBAN	25304.3	yes	2	yes	no	no	no	no
ID12108	54	male	Town	24212.1	yes	2	yes	yes	yes	no	no
ID12108	66	female	Town	59803.9	yes	0	no	yes	yes	no	no

TABLE 7.5 Sample Mail Campaign Data.

your focus is to determine likely responders to a direct mail campaign given the following strategies:

1. a new product, is a "Personal Equity Plan" (PEP).

2. training data include records kept about how previous customers responded and bought the product.

3. in this case the target class is "PEP" with a binary value.

4. build a model and apply it to new data (a customer list) in which the value of the class attribute is not available.

8. For the following table:

Tuple	Items
T1	Bread, jelly, peanut butter
T2	Bread, peanut butter
T3	Bread, milk, peanut butter
T4	Soda, bread
T5	Soda, milk

TABLE 7.6 Transaction Data Set.

(a) Intuitively list the frequent item sets

(b) What is the support count for:

a. {bread, peanut butter}

b. {soda, milk}

(c) What is the support for:

a. {bread, peanut butter}

b. {soda, milk}

9. Complete the following table:

Association rule	Support	Confidence
Bread → peanut butter		
Peanut butter → bread		
Soda → bread		
Peanut butter → jelly		
Jelly → peanut butter		
Jelly → milk		

10. For the following database:

Tuple	Items
10	A, B, D
20	B, C, E
30	A, B, C, E
40	B, E

FIGURE 7.12 Hypothetical Transaction Database

Derive the results of the apriori algorithm on the database.

11. When constructing a classification tree, how can overfitting be accomplished?

12. How can you determine the correct size of a classification tree?

13. Outline the rule postpruning algorithm and compare it to tree pruning.

14. What problems are appropriate for decision tree learning?

15. Which are the main features of decision tree learning?

16. Consider the following table of purchases at a local office supply store:

Trans ID	Cust ID	Date	Item	Qty
1	17	3/1/12	HD	1
1	17	3/1/12	Flash drive	1
1	17	3/1/12	Software	1
2	4	3/8/12	Flash drive	2
2	4	3/8/12	Software	1
2	4	3/8/12	Paper	4
2	4	3/8/12	Ink cart	2
2	4	3/8/12	HD	1
3	9	3/15/12	Ink cart	1
3	9	3/15/12	Flash drive	1
3	9	3/15/12	Software	1
4	11	3/22/12	Ink cart	1
4	11	3/22/12	HD	2
4	11	3/22/12	Paper	1

TABLE 7.8 Office Supply Store Purchase Transactions.

(a) Find the frequent item sets with minsup = 0.90 and then find association rules with minconf = 0.90.

(b) Find the frequent item sets with minsup = 0.10 and then find association rules with minconf = 0.90.

17. Consider the table in Problem 16. Find all association rules that indicate the likelihood of items being purchased on the same date by the same customer, with minsup = 0.10 and minconf = 0.70.

18. Operationally define support and confidence for an association rule.

19. Can we use association rules for prediction?

CHAPTER **8**

CLUSTER VALIDITY

In This Chapter

8.1 INTRODUCTION

Once the final clustering has been obtained the major step to cluster analysis begins. Are the points within clusters representative in reality? Do the between cluster distances reflect the actual situation in reality? What is needed is an objective means, not subjective, to determine the accuracy and validity of the final clustering chosen in a study. *Internal indices* measure the inter-cluster validity and *external indices* measure the intra-cluster validity. Additionally, when real data sets are readily accessible then the validity indices can be subjected to statistical testing. *Monte Carlo analysis*, studies based upon the application being run on randomly generated data, can be employed when real data sets are not accessible.

Besides concentrating on validation, a cluster analysis should emphasize the *stability* of the results, as is needed for nearest neighbor and furthest neighbor hierarchical clustering. Stability of the results can be accomplished by adding new patterns or removing some features and repeating the clustering method. The resulting clustering and the modified clustering will then need to be compared.

Sometimes the investigator needs to determine which of a set of clustering best fits the data. Indices of this type are referred to as *relative indices*.

8.2 STATISTICAL TESTS

Hypothesis testing is the third estimation procedure presented, in addition to point estimation and interval estimation. In hypothesis testing, we pose two mutually exclusive statements, or *hypotheses*, about a population parameter, then develop a *decision rule*, which is a procedure based on one or more statistics of a random sample that will determine which of the two hypotheses we should accept as true.

We will use examples to illustrate the basic process for testing hypotheses. In hypothesis testing, two mutually exclusive hypotheses are given. The *null hypothesis* (designated H_0) states the *status quo*, the condition the researcher usually wishes to disprove, while the *alternative hypothesis* (H_a) is generally what the researcher wishes to show, that is, what will be true if conditions have changed.

Example 1

Suppose that drug A has been used to treat a particular disease. A group of medical researchers develops drug B, which they hope will be more useful in combating the disease, but before it can be commonly used, the researchers must demonstrate that their new medicine is more effective than drug A. They will conduct an experiment to evaluate the two treatments, and will perform a *hypothesis test*.

Solution:

In the drug example, the hypotheses are these:

H_0: Drug A is at least as effective as drug B.

H_a: Drug B is more effective than drug A.

Hypothesis testing is a conservative process, based on the presumption that H_0 is true. The researcher hopes to reject H_0 and accept H_a on the basis of experimental (sample) evidence, but if H_0 is not decisively shown to be true, then the researcher fails to reject H_0, and no definite conclusion is reached.

What is needed is a procedure for deciding between the null and alternative hypotheses, a well-defined rule by which the decision, based on sample information, can be made. Consider another example.

Example 2

A manufacturer of TVs has been producing a 19-inch tube that has an expected life of 3,100 hours, with a standard deviation of 450 hours. An engineer suggests a modification that she believes will increase the lifetimes of such TVs. To test her hypothesis, she tests a sample of 100 TVs built with the modification, and finds that the sample has a mean lifetime of 3,225 hours. Is this convincing evidence that the modification should be incorporated in all the company's picture tubes? Does it demonstrably extend expected tube life?

Solution:

The hypotheses being tested are:

H_0: The modification causes no improvement.

H_a: The modification extends the TV life.

Rephrasing these in terms of a parameter, the population mean,

$$H_0: \mu \le 3100$$

$$H_a: \mu > 3100$$

where μ is the expected lifetime of picture tubes incorporating the modification. Recall that the expected life of TVs without the modification was 3,100 hours.

The engineer must decide whether to reject H_0 and accept H_a (conclude that the modification is effective) or to not reject H_0 (conclude that lifetimes of the TVs incorporating the modification are no longer than those of the original design).

Hypothesis testing leads to four possible situations, as shown in Figure 8.1.

Decision

	Do not Reject the Null Hypothesis	Reject the Null Hypothesis
Null Hypothesis True	Correct	Type I error
Null Hypothesis False	Type II error	Correct

State of Reality

FIGURE 8.1 Outcomes of a Hypothesis Test.

Failing to reject H_0 when H_0 is true and rejecting H_0 when it is false are correct decisions. The other two possibilities represent errors. Rejecting the null hypothesis when it is, in fact, true is a *Type I error*, while failing to reject a false null hypothesis is a *Type II error*. In Example 2, a Type I error is committed if the engineer concludes that the modification is effective in prolonging tube life when it is not, and a Type II error is committed if the engineer fails to conclude that the modification is effective when, in fact, it is not.

The Type I error is generally more serious. In Example 2, it might result in spending time and money to incorporate an ineffective modification, so hypothesis tests are constructed to control the probability of a Type I error. This probability is indicated by α. In solving Example 2, we construct a rule by which we can decide if μ is greater than 3,100. We have seen the utility of estimating μ with the sample mean $\overline{X}$, so we will be convinced to reject the null hypothesis and accept the alternative if $\overline{X}$ is big enough, but how big is big enough? Because we also want to control $\alpha = P(\text{Type I error})$, the probability that H_0 is rejected when it is true, consider the sampling distribution of $\overline{X}$ if H_0 is true, as shown in Figure 8.2.

If H_0 is true, then $\mu = 3,100$, and we know from the central limit theorem that the sampling distribution of $\overline{X}$ is approximately normal with mean 3,100 and standard deviation $450/\sqrt{100}$. Choose a value for α, say 5%, and

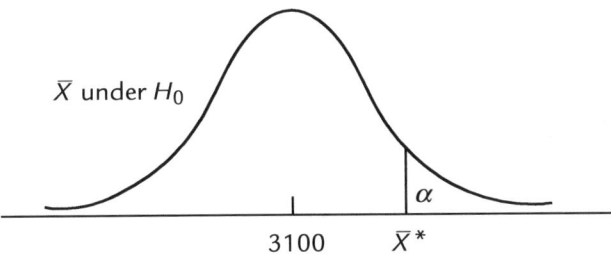

$\bar{X}$ under H_0

α

3100 $\bar{X}^*$

FIGURE 8.2 Sampling Distribution of $\bar{X}$ Given that H_0 is True.

cut off an area of size α in the upper tail of the graph of this distribution. Consider the value $\bar{X}^*$ that marks off its lower edge.

If H_0 is true, the probability that $\bar{X}$ will be greater than $\bar{X}^*$ is α, which we have chosen to be only 5%, so if the observed value of $\bar{X}$ is greater than $\bar{X}^*$, we can reject H_0 and accept H_a, concluding that the modification extends tube life, with only 5% probability of making a Type I error. We need only find the value of $\bar{X}^*$:

$$\alpha = P(\text{Type I error}) = P(\text{Reject } H_0 \mid H_0)$$

$$= P(\bar{X} > \bar{X}^* \mid H_0) = 5\% = 0.05$$

$$\rightarrow P\left(\frac{\bar{X} - 3100}{450/\sqrt{100}} > \frac{\bar{X}^* - 3100}{450/\sqrt{100}} \,\middle|\, \mu = 3100 \right) = 0.05$$

$$\rightarrow P\left(Z > \frac{\bar{X}^* - 3100}{450/\sqrt{100}} \right) = 0.05$$

$$\rightarrow P\left(0 < Z < \frac{\bar{X}^* - 3100}{450/\sqrt{100}} \right) = 0.45$$

$$\rightarrow \frac{\bar{X}^* - 3100}{450/\sqrt{100}} = 1.645$$

$$\rightarrow \bar{X}^* = 3100 + 1.645 \frac{450}{\sqrt{100}} = 3174.025.$$

This value completes construction of the decision rule:

If the sample mean $\bar{X}$ is greater than $\bar{X}^* \cong 3174$, then we reject H_0 and accept H_a. If $\bar{X} < (\bar{X}^* \cong 3174)$, we cannot reject H_a.

The probability α is called the *significance level* of the test, and $\overline{X}^*$ is the *critical value*. Because our conclusion will be determined by the value of $\overline{X}$, the sample mean is called the *test statistic*. The interval above the critical value in this *upper-tail test* is called the *region of rejection*, because we reject H_0 when $\overline{X}$ falls in this region, while the interval below $\overline{X}^*$ is called the *region of acceptance*, or the area where we fail to reject.

In her examination of the 100 tubes made with the modification, the engineer found that $\overline{X} = 3225$, which is greater than the critical value of 3174. According to the decision rule, she can reject H_0 at the 5% level of significance and conclude that the modification does extend the expected life of picture tubes.

In general, suppose we take a sample of size $n > 30$ to perform an upper-tail test of the form

$$H_0: \mu \leq \mu_0$$
$$H_a: \mu > \mu_0.$$

We choose the significance level α, and must find the critical value $\overline{X}^*$ around which to build the decision rule. If H_0 is true, the distribution of the sample mean $\overline{X}$ is approximately normal with mean μ and standard deviation $\sigma/\sqrt{n}$, where σ is the population standard deviation and n is the sample size. Then

$$\alpha = P(\text{Type I error}) = P(\text{Reject } H_0 \mid H_0) = P(\overline{X} > \overline{X}^*)$$

$$P\left(\frac{\overline{X} - \mu_0}{\sigma/\sqrt{n}} > \frac{\overline{X}^* - \mu_0}{\sigma/\sqrt{n}} \middle| H_0\right) = \alpha \rightarrow P\left(Z > \frac{\overline{X}^* - \mu_0}{\sigma/\sqrt{n}}\right) = \alpha$$

$$\rightarrow \frac{\overline{X}^* - \mu_0}{\sigma/\sqrt{n}} = z_\alpha.$$

Where z_α, called the *critical normal deviate*, is chosen so that $P(Z > z_\alpha) = \alpha$. Then $\overline{X}^* = \mu_0 + z_0\left(\sigma/\sqrt{n}\right)$.

If $\overline{X} > \overline{X}^*$, we reject H_0 and accept H_a. If $\overline{X} < \overline{X}^*$, we fail to reject H_0, and no conclusion is reached, as shown in Figure 8.3.

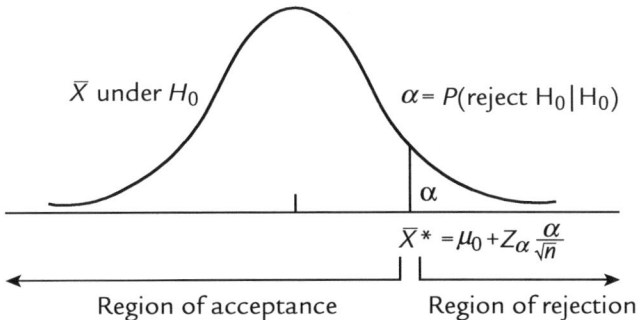

FIGURE 8.3 An Upper-Tail Hypothesis Test for μ.

If the population standard deviation is unknown, as it often is in sampling situations, the standard sample standard deviation is used instead of σ. Then $\overline{X}^{*} = \mu_{0} + z_{0}\left(s/\sqrt{n}\right)$.

The decision rule for this kind of hypothesis test can be phrased in two different but equivalent ways. First, notice that if $\overline{X} > \overline{X}^{*} = \mu_{0} + z_{0}\left(\sigma/\sqrt{n}\right)$, then $\dfrac{\overline{X} - \mu_{0}}{\sigma/\sqrt{n}} > z_{\alpha}$. That is, we can compare the value of the normal deviate $\dfrac{\overline{X} - \mu_{0}}{\sigma/\sqrt{n}}$ to the critical normal deviate, as shown in Figure 8.4.

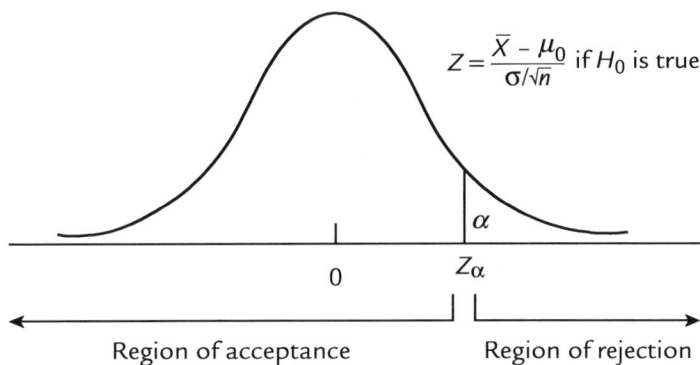

FIGURE 8.4 Regions of Rejection and Acceptance [Fail to Reject].

If $\dfrac{\overline{X} - \mu_{0}}{\sigma/\sqrt{n}} > z_{\alpha}$, reject H$_0$ and accept H$_a$.

If $\dfrac{\overline{X}-\mu_0}{\sigma/\sqrt{n}} < z_\alpha$, do not reject H_0.

In Example 2, $\alpha = 5\%$ so $z_\alpha = 1.645$.

$$\frac{\overline{X}-\mu_0}{\sigma/\sqrt{n}} = \frac{3225-3100}{450/\sqrt{100}} = 2.78 > 1.645.$$

As before, we conclude that the engineer should reject the null hypothesis and conclude that the modification does increase expected tube life.

The second way to phrase the decision rule compares the area under the graph of the distribution of $\overline{X}$ when H_0 is true above the observed value of $\overline{X}$ to the significance level of α. If this area is less than α, then $\overline{X}$ itself must be above the critical value $\overline{X}^*$, and we reject H_0; if the area above the observed value of $\overline{X}$ is greater than α, then $\overline{X}$ must be below $\overline{X}^*$, and we cannot reject H_0, as shown in Figure 8.5. The area above $\overline{X}_o$, the observed value of $\overline{X}$, is $P(\overline{X} > \overline{X}_o | H_0)$.

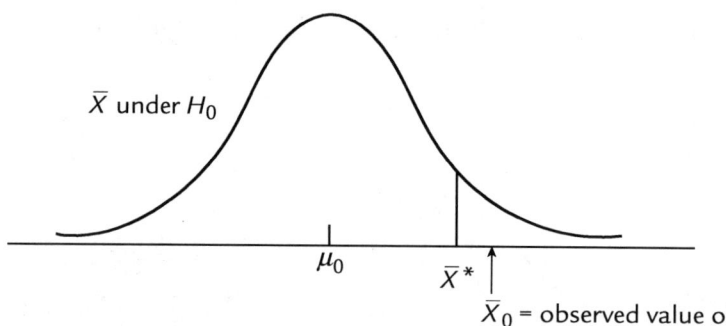

$\overline{X}$ under H_0

μ_0

$\overline{X}^*$

$\overline{X}_0$ = observed value of $\overline{X}$

FIGURE 8.5 One-Tail Probability.

Again looking at Example 2, $\alpha = 5\% = 0.05$, and the observed value of $\overline{X}$ was $\overline{X}_o = 3225$. Then

$$P(\overline{X} > 3225 | H_0) = P\left(\frac{\overline{X}-3100}{450/\sqrt{100}} > \frac{3225-3100}{450/\sqrt{100}} \middle| H_0 \right)$$

$$= P(Z > 22.78) = 0.5000 - 0.4973 = 0.0027.$$

$$0.0027 < 0.05 = \alpha, \text{ so we reject } H_0.$$

The value $P\left(\overline{X} > \overline{X}_o | H_0\right)$ is sometimes called a *one-tail probability*. Note that this is the probability of a value of $\overline{X}$ at least as extreme as the observed value that will occur if H_0 is true.

Other forms of hypothesis tests are possible. We can perform *lower-tail* or *two-tail tests* with the population mean μ, as well as tests involving other population parameters. All statistical tests of hypotheses, however, will contain these elements:

- A formal statement of the null and alternative hypotheses, H_0 and H_a

- A test statistic and its sampling distribution

- A chosen level of significance, α

- A decision rule that defines the critical value(s) of the test statistic and the regions of acceptance and rejection

- A random sample from which to obtain the observed value of the test statistic

We have seen that the decision rule of a hypothesis test is developed to correspond to our choice of the significance level, the probability of a Type I error. We determine, and keep small, the probability of rejecting the null hypothesis when it is true. Suppose, however, that we fail to reject H_0. How much confidence can we have that we have not committed a Type II error? We must investigate β, the probability of failing to reject the null hypothesis when it is false.

Reconsider Example 2, in which the engineer hopes to demonstrate, using a sample size of 100, that a modification to her company's 19-inch picture tubes will increase their expected life beyond 3,100 hours. The population standard deviation is 450 hours and the hypotheses, to be tested at the 5% significance level, are these:

$$H_0: \mu \leq 3100$$
$$H_a: \mu > 3100.$$

The critical value of this test is

$$\overline{X}^* = 3100 + 1.645 \frac{450}{\sqrt{100}} \cong 3174,$$

and the decision rule is:

$$\text{If } \overline{X} > 3174, \text{ reject } H_0 \text{ and accept } H_a;$$
$$\text{If } \overline{X} < 3174, \text{ do not reject } H_0.$$

Suppose that the expected lifetime of tubes incorporating the modification is 3,200 hours. Then $\mu = 3200$, H_0 is false, and H_a is true. In this situation, the sampling distribution of $\overline{X}$ is approximately normal with mean 3200 and standard deviation $450/\sqrt{100}$, and the probability of a Type II error, or failing to conclude that H_0 is false even though H_a is true (since $\mu = 3200$), is:

$$\beta = P(\text{Type II error}) = P(\overline{X} < 3174 | H_z \text{ with } \mu = 3200)$$

$$= P\left(\frac{\overline{X} - 3200}{450/\sqrt{100}} < \frac{3174 - 3200}{450/\sqrt{100}} \middle| H_z \text{ with } \mu = 3200 \right)$$

$$= P(Z < -0.58) = 0.5000 - 0.2190 = 0.2810.$$

That is, if the true mean lifetime of tubes incorporating the new process is 3,200 hours, the probability that our test will nonetheless fail to reject H_0 is 28.10%. This probability is represented graphically in Figure 8.6 as the area under the curve of the distribution of $\overline{X}$ if $\mu = 3200$. Note that 3200 is to the left of $\overline{X}^* = 3174$.

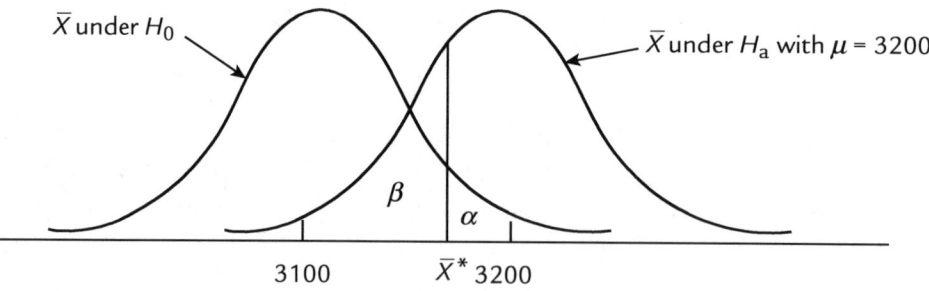

$\overline{X}$ under H_0 $\overline{X}$ under H_a with $\mu = 3200$

β

α

3100 $\overline{X}^*$ 3200

FIGURE 8.6 Comparison of H_0 Distribution to an Alternative Distribution.

In general, for an upper-tail test of this kind,

$$\beta = P(\overline{X} < \overline{X}^* | H_a \text{ with } \mu = \mu_a) = P\left(Z < \frac{\overline{X}^* - \mu_a}{\sigma/\sqrt{n}} \right),$$

where μ_a is a possible value of μ for which H_0 is false and H_a is true.

It is important to observe that the value of β depends on the true value of the population mean μ, which is also the center of the distribution of $\overline{X}$. If μ is, in fact 3125,

$$\beta = P(\text{Type II error}) = P(\overline{X} < 3174 | H_z \text{ with } \mu = 3125)$$

$$= P\left(\frac{\overline{X} - 3125}{450/\sqrt{100}} < \frac{3174 - 3125}{450/\sqrt{100}} \middle| H_z \text{ with } \mu = 3125 \right)$$

$$= P(Z < -1.00) = 0.5000 \pm 0.3621 = 0.8621.$$

If the true mean is 3,125, the probability that we will mistakenly fail to reject H_0 is 86.21%, as shown in Figure 8.7.

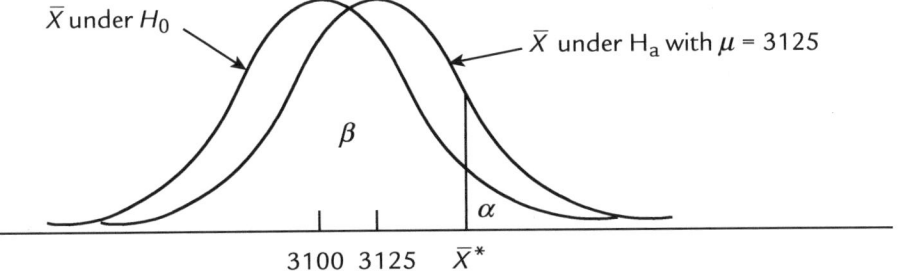

FIGURE 8.7 Failure to Reject H_0.

There is a value of β corresponding to every possible value of μ for which H_0 is false. Because there are an infinite number of such values, it would be futile to attempt to calculate them all. We can, however, represent graphically the relationship between $P(\text{Accept } H_0)$ and all possible values of μ.

If $\mu = 3100$, then H_0 is true, and $P(\text{Accept } H_0 | H_0) = 1 - P(\text{Reject } H_0 | H_0) = 1 - \alpha = 0.95$. Calculations like these fill in the rest of Table 8.1, which shows the representative values of μ.

μ	P(Accept H$_0$)
3100	0.9500
3125	0.8621
3150	0.7019
3174	0.5000
3200	0.2810
3225	0.1292
3250	0.0455

TABLE 8.1 Representative Values of μ.

Note that if $\mu = \overline{X}^*$, that is, if the population mean is equal to the critical value, the value of β is 0.5000.

From these values, we draw Figure 8.8 relating P(Accept H_0) to possible values of μ. It is called the *operating characteristic curve* or *OC curve* for the test, and at each possible value of μ except 3,100 (where H_0 is true), the height of the graph is the probability of a Type II error, β. The OC curve always begins at the point $(\mu_0, 1-\alpha)$, where μ_0, is the null hypothesis value of μ, and passes through the point $(\overline{X}^*, 0.5)$.

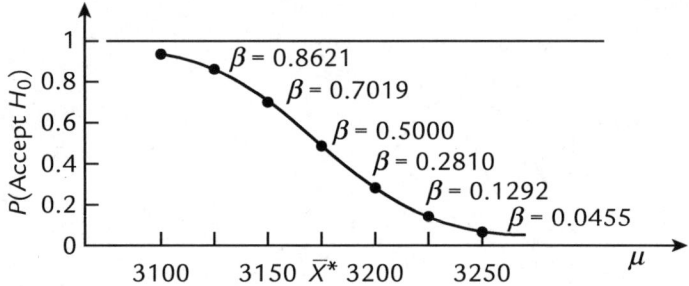

FIGURE 8.8 Operating Characteristic Curve.

Used as often as the operating characteristic curve is the *power curve*, which plots $1 - \beta = $ P(Reject H_0) against possible values of μ, as shown in Figure 8.9.

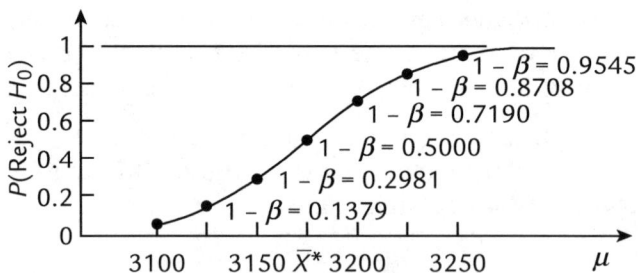

FIGURE 8.9 Power Curve.

Note that the curve begins at the point (μ_0, α), also passes through $(\overline{X}^*, 0.5)$, and contains the same information as the operating characteristic curve.

The graphs in Figures 8.8 and 8.9 tell us the probabilities of accepting or rejecting the null hypothesis for all possible values of μ; they tell us how likely it is that our test will distinguish between μ_0 (3,100 in Example 2) and other possible values of μ.

These curves can also be used to illustrate how a test is affected by changes in significance level or sample size. For example, in an upper-tail test for μ of the type we have been considering, an increase in the significance level α will shift the starting point of the OC curve down and that of the power curve up, and will decrease the critical normal deviate z_α so that $\overline{X}^* = \mu_0 + z_\alpha \sigma/\sqrt{n}$ also decreases.

In Figure 8.10, the OC curve is moved down and the power curve up. That is, if the significance level is increased, the probability of a Type II error, β, is decreased. Conversely, decreasing α increases β. Again, we encounter a trade-off, like that between certainty and precision in confidence intervals.

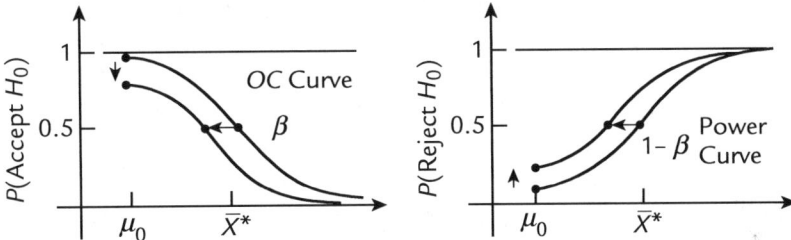

FIGURE 8.10 Effects on the OC and Power Curves of Increasing α.

With confidence intervals, we found that we could improve certainty without degrading precision and, vice versa, by collecting more information; that is, by increasing the sample size n. In our upper-tail hypothesis test for μ, increasing n will decrease the standard deviation of $\overline{X}$, $\sigma/\sqrt{n}$, and will reduce the critical value $\overline{X}^* = \mu_0 + z_\alpha \dfrac{\sigma}{\sqrt{n}}$. The starting points of the two curves, $(\mu_0, 1-\alpha)$ and (μ_0, α), will not be affected, so the curves will become steeper as shown in Figure 8.10. Except in a small region near μ_0, β will be reduced without degrading α; for most values of μ above μ_0, the test with larger n is more likely to be accurate.

In general, if the OC and power curves of one test are steeper than those of another, we say that the first test is more *powerful*, or better able to distinguish between μ_0 and other possible values of μ. We can increase the power of a test, its ability to distinguish between H_0 and H_a, by enlarging the sample.

8.3 MONTE CARLO ANALYSIS

The methodology of simulation, initially used by operations research, was created by John von Neumann and Stanislaw Ulam in the late 1940s. Since the advent of digital computing, simulation has been used for countless business applications. Essentially, simulation is the use of a system model that has the desired characteristics of reality in order to reproduce the essence of the actual operations. The Monte Carlo method is simulation by sampling techniques. It involves determining the probability distribution of the variable under consideration and then sampling from this distribution by means of random numbers to obtain data. In effect, a set of random numbers is used to generate a set of values that have the same distributional characteristics as the actual experience it is designed to simulate.

In order to perform a Monte Carlo Analysis, we need to examine *random number generation*. One of the key steps in performing a statistical analysis for a simulated study is often to have a routine that generates random values for variables with a specified random distribution; for example, exponential and normal. This is done in two steps. First, a sequence of random numbers distributed uniformly between 0 and 1 is obtained. Then the sequence is transformed to produce random values satisfying the desired distribution. The first step is called *random number generation* and the second, is *random variate generation.*

Most random number generators are defined as a recursive relation in which the next number in the sequence is a function of the last one or two numbers. For example,

$$X_n = \left(5X_{n-1} \right) \bmod 16$$

defines, starting with $X_0 = 5$, the random sequence:

5, 10, 3, 0, 1, 6, 15, 12, 13, 22, 11, 8, 9, 14, 7, 4, 5.

Note that $X_1 = (5(5) + 1)\bmod 16 = 26 \bmod 16 = 10$. For this random number generator, all the X_i values are integers between 0 and 15. Also take note that the sequence regenerates 5. The fact that the number 5 is regenerated means the sequence can be regenerated any time provided the starting value X_0 is given. The value that is used to begin the sequence is called the *seed.*

Given the seed, we can tell with a probability of 1 the numbers in the sequence that will be generated, or we say the function representing the random number generator is *deterministic.* Yet the numbers to be discussed are random in the sense that they would pass statistical tests for randomness. These numbers are, therefore, only partially random and are called *pseudorandom numbers.*

In our example, only 16 numbers are unique within the sequence, and then these 16 numbers are repeated. In other words, this random number generator has a *cycle length* of 16. If a generator does not repeat an initial part of the sequence, then that part is referred to as the *tail.* The *period* of the generator is the sum of the tail length and the cycle length.

The desirable properties for a random number generator are:

- The related software should execute efficiently.

- The sequence should possess a large period.

- The numerical values in the sequence should be independent and uniformly distributed.

The last property requires a battery of statistical tests to validate randomness.

Some of the random number generators commonly used include:

- Linear congruential generators

- Extended Fibonacci generators

- Combined generators

A description of each of these approaches follows.

The residues of successive powers of a number possess good randomness properties, where the nth number in the sequence is found by dividing the nth power of an integer a by another integer m and keeping the remainder. That is,

$$X_n = aX_{n-1} \bmod(m).$$

The parameter a is called the *multiplier* and m is called the *modulus.* Such generators are referred to as *Linear Congruential Generators* (LCGs).

In general, the choice of a, b, and m affects the period and autocorrelation in the resultant sequence. Generally,

- The modulus m should be large.

- For $\bmod(m)$ computations to be efficient, m should be a power of 2.

- If b is nonzero, the maximum possible period m is obtained when:

  - m and b are relatively prime (they have no common factors other than 1).

  - Every prime number that is a factor of m is also a factor of $a - 1$.

  - $a - 1$ is a multiple of 4.

  Notice that all of these conditions are met if $m = 2^k$, $a = 4c + 1$, and b is odd. It is assumed that c, b, and k are positive integers.

  A *multiplicative LCG* is of the form:

  $$X_n = aX_{n-1} \bmod(m).$$

  Multiplicative LCGs are more efficient than mixed LCGs. Additional efficiency can be obtained by choosing m to be a power of 2. However, the maximum possible period for a multiplicative LCG with modulus $m = 2^k$ is only one-fourth the full period possible, that is, 2^{k-2}. The latter period is achieved if the multiplier a is of the form $8i \pm 3$ and the initial seed is an odd integer.

  A solution to the small period problem is to use a modulus m that is a prime number. In this case, with a proper choice of the multiplier a, it is possible to get a period of $m - 1$, which is almost equal to the maximum possible length m. Notice that unlike a mixed LCG, X_n obtained from a multiplicative LCG can never be 0 if m is prime. The values of X_n lie between 1 and $m - 1$, and any multiplicative LCG with a period of $m - 1$ is called a *full period generator.*

  Not all values of the multiplier are equally good. It can be shown that a multiplicative LCG will be a full period generator if and only if the multiplier a is a *primitive root* of the modulus m. By definition, a is a primitive root of m if and only if $a^n \bmod(m) \neq 1$ for $n = 1, 2,\ldots, m - 2$.

  One of the important cautions in implementing LCGs is that the properties are guaranteed only if the computations are done exactly and without any rounding errors. Another problem in implementing LCGs is that the product aX_{n-1} can exceed the largest integer allowed on the computer system, causing integer overflow.

  A Fibonacci sequence $\{X_n\}$ is generated by the following relationship:

  $$X_n = X_{n-1} + X_{n-2}.$$

Random numbers can be generated using the Fibonacci sequence in this way:

$$X_n = \left(X_{n-1} + X_{n-2}\right)\mathrm{mod}(m).$$

However, this sequence does not possess good randomness properties as it has a high serial correlation. But the following relationship passes most statistical tests for randomness:

$$X_n = \left(X_{n-5} + X_{n-17}\right)\mathrm{mod}(2^k).$$

It is possible to combine two or more random generators to produce a "better" generator. The following are three such techniques:

1. Adding random numbers obtained by two or more generators.

Note that this modification will sometimes increase the period and randomness (L'Ecuyer).[1]

2. Exclusive: or random numbers obtained by two or more generators (Santha and Vazirani).[2]

3. Shuffle: shuffling is when one sequence is used as an index to decide which of several numbers generated by a second sequence should be returned (Marsaglia and Bray).[3] Algorithm M by Marsaglia and Bray uses an array of size 100 that is filled with random numbers from a random sequence X_n. To generate a random number, generate a new Y_n (between 0 and $m-1$) and scale it to obtain an index $I = 1 + 100\,Y_n/m$. The value in the ith element of the array is returned as the next random number. A new value of X_n is generated and stored in the ith location. One problem with shuffling is that it is not easy to skip a long subsequence as is often required in applications.

---

[1] L'Ecuyer, P. (1988). Efficient and portable combined random number generators. *Communications of the ACM, 31,* 742-749 and 774.

[2] Santha, M., & Vazirani, U. V. (1984). Generating quasi-random sequences from slightly random sources. *Proceedings 25th Annual Symposium on Foundations of Computer Science,* 434-440.

[3] Marsaglia, G., & Bray, T. A. (1964). A convenient method for generating normal variables. *SIAM Review, 6,* 260-264.

The following guidelines should be taken into consideration when selecting a seed for a random number generator:

- Do not use 0. This choice of value would make a multiplicative LCG stick at 0.

- Avoid even values. If you are using a multiplicative LCG, the seed should be odd.

- Do not subdivide one stream. Using a single stream for all variables is a common mistake that can result in a strong correlation between the variables.

- Use nonoverlapping streams. If the streams overlap, there will be a correlation between the streams, and the resulting sequences will not be independent.

- Reuse seeds in successive replications.

- Do not use random seeds. This causes two problems: First, the results are difficult to reproduce; and, second, it is not possible to guarantee that the multiple streams will not overlap.

The following are some common misconceptions about random number generation:

- A complex set of operations leads to a random number.

- A single test such as the chi-square test is sufficient to test the goodness of fit for a random number generator.

Note that the sequence 0, 1, 0, 1, 0, 1,..., 0, 1 will pass the chi-square test, but clearly the sequence is not random (a pattern is present).

- Random numbers are unpredictable. LCGs are deterministic.

- Some seeds are better than others. In general, generators whose period or randomness depends upon the seed should not be used, because an unsuspecting user may not remember to follow the guidelines.

- Accurate implementation is not important. The period and randomness properties of generators are guaranteed only if the generation formula is accurately implemented without any overflow or truncation.

- Bits of successive words generated by a random number generator are equally randomly distributed. There are a number of methods to generate nonuniform variables. For a particular distribution, one method may be more efficient than the others. In this chapter, some of the commonly used methods are described.

We will look at four methods for generating nonuniform variables: inverse transformation, rejection, composition, and convolution. These methods assume that we have already generated a sequence of random numbers distributed uniformly between 0 and 1.

Suppose we want to generate random values from A to B from a uniform distribution defined over [A, B]. We recall that the cumulative distribution function (CDF) in this case is monotonically increasing and continuous over the interval and is defined as:

$$P(x) = \frac{x - A}{B - A}, \text{ for } A \leq x \leq B.$$

Then in the *inverse method* of set r, a uniform value defined on [0, 1], equal to P(x), giving:

$$r = P(x) = \frac{x - A}{B - A}.$$

Because the cumulative distribution function is one to one, onto and continuous on [A, B], then the inverse function for P(x) exists. This "inversing" of variables gives the inverse method its name.

$$r = \frac{x - A}{B - A}$$

then

$$x = A + r(B - A).$$

The final equation is easy to use as a process generator for the uniform distribution. Given values for the parameters (A and B) and a random number on [0, 1], a sampled value (a variate) is obtained by substituting the appropriate values into the last equation. For example, if the distribution of interest is uniformly distributed with a lower limit of 5 and an upper limit of 10 and the random number 0.75 is generated, the sampled value would be

$$x = 5 + 0.75(10 - 5)$$

$$= 8.75.$$

Consider the need to sample a value from the negative exponential distribution. The first step is using the inverse transformation method to obtain the cumulative distribution function, which we know is:

$$P(x) = 1 - e^{-\lambda x}.$$

Next, the uniform random number r is set equal to $P(x)$, and the expression is manipulated to express x as a function of r:

$$r = P(x) = 1 - e^{-\lambda x}$$

$$e^{-\lambda x} = 1 - r.$$

Taking natural logarithms of each side, we get

$$-\lambda x = ln(1 - r)$$

$$x = \frac{-ln(1 - r)}{\lambda}.$$

The latter equation is our random variate generator for the negative exponential distribution. Given a mean number of occurrences and a uniform random number, the number of intervals before the next occurrence is easily simulated. Keep in mind, however, that the sampled value is measured in intervals and must usually be transformed into minutes, hours, or whatever the appropriate unit of measurement may be before it becomes useful in an application. For example, if $\lambda = 1$ occurrence 5 minutes and $r = 0.40$, the sampled value is:

$$X = \frac{-ln(0.6)}{1} = 0.52.$$

Measured in minutes, this is (0.51 intervals)*(5 minutes/interval) = 2.55 minutes.

A *rejection technique* can be used if another density function, the Probability Density Function (pdf), $g(x)$ exists so that $cg(x)$ *majorizes* the density function $f(x)$; that is, $cg(x) \geq f(x)$ for all values of x. If such a function can be found, then the following steps can be used to generate the random variate x with density $f(x)$:

1. Generate x with pdf $g(x)$.

2. Generate y uniform on $[0, cg(x)]$.

3. If $y \leq f(x)$, then output x and return. Otherwise, repeat from 1.

The algorithm continues *rejecting* the random variates x and y until the condition $y \leq f(x)$ is satisfied; hence, the name for the method.

Consider the problem of sampling values from a normally distributed random distribution. The inverse function does not exist; therefore the inverse transformation method does not meet our needs. However, the rejection method does solve the problem. In fact, this method works for any continuous distribution.

First, enclose the continuous distribution in a rectangle, letting a rectangular distribution majorize the continuous distribution. Because the standard normal distribution is asyptomatic to the z-axis, it is necessary to truncate the distribution at some reasonable point, as shown in Figure 8.11.

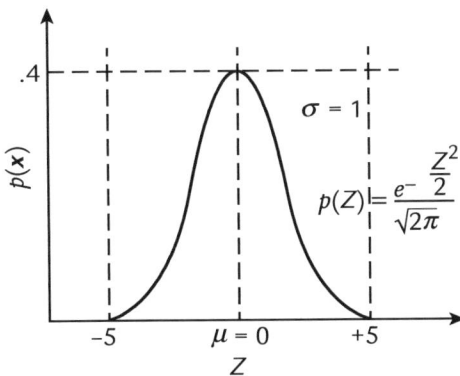

FIGURE 8.11 The Majorizing of the Standard Normal Distribution.

Points that are five standard deviations from the mean are chosen as the truncation points. The maximum height of the curve is 0.3989. Consequently, a rectangle with a height of 0.40 includes the entire distribution.

The next step is to randomly select a Z value between +5 and –5.

$$Z = -5 + 10r_1.$$

After the Z value has been randomly selected, it is inserted in the equation for the standard normal distribution to determine if the height of the curve is at that point. Then a second random number is generated. It is transformed, using the uniform process generator, so that it randomly selects a point along the height of the rectangle.

$$H = 0.4r_2.$$

Now comes the rejection part of the algorithm. If the value of H falls below the computed height of the curve at the randomly selected Z value, the Z value is treated as if it comes from the standard normal distribution. If, however, the value of H falls above the curve, the Z value under consideration is rejected and the process is repeated until an appropriate value is found.

The *composition method* can be used if the desired Cumulative Distribution Function (CDF) can be expressed as a weighted sum of n other CDFs. That is,

$$F(x) = \sum_{i=1}^{n} p_i F_i(x)$$

where, $p_i \geq 0, \sum_{i=1}^{n} p_i = 1,$ and the $F_i(x)$'s are distribution functions. The technique can also be used if the density function $f(x)$ can be expressed as a similar weighted sum of n other density functions.

In either case, the steps to generate x are as follows:

1. Generate a random integer where
$$P(I = i) - p_i$$

This can easily be done using the inverse transformation method.

2. Generate x with the ith $f_i(x)$ and return.

The pdf for a Laplace distribution is a composition of two exponential pdfs. Figure 8.12 illustrates the Laplace distribution given by the pdf:

$$f(x) = \frac{1}{2a} e^{\frac{-|x|}{a}}$$

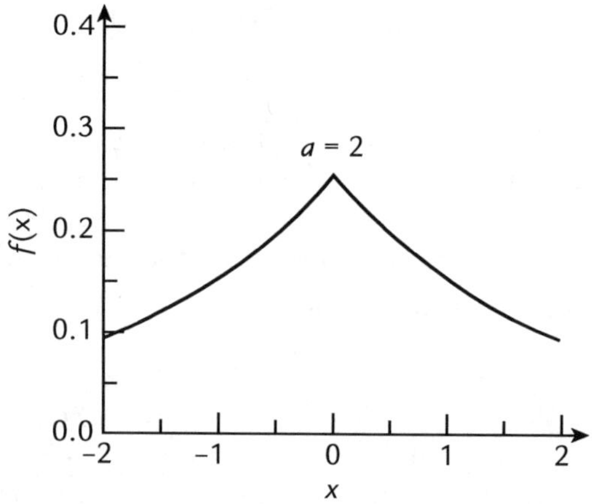

FIGURE 8.12 Laplace Density Function.

The probability of x being positive is half, and the probability of x being negative is half.

By the composition technique, Laplace variates can be generated in the following way:

1. Generate two uniform $(0, 1)$ values r_1 and r_2.

2. If $r_1 < 0.5$, return $x = -aln(r_2)$; otherwise, return $x = aln(r_2)$.

The *convolution method* can be used if the random variable x can be expressed as a sum of n random variables $y_1, y_2, \ldots, y_n$ that can be easily generated; that is,

$$x = y_1 + y_2 + \ldots + y_n.$$

Notice the difference between composition and convolution. The former method is used when the pdf or CDF can be expressed as a sum of other pdfs or CDFs. The latter technique is used when the random variable itself can be expressed as a sum of other random variables. The following list of examples illustrate this summation process:

- A binomial variate with parameters n and p is a sum of n Bernoulli variates with success probability p.

- The chi-square distribution with υ degrees of freedom is a sum of squares of υ unit normal $N(0,1)$ variates.

- The sum of a "large" number of variates from any distribution has a normal distribution.

- The sum of two uniform variates has a triangular density.

Reconsider the normal distribution. Begin by obtaining 12 uniformly distributed variates, computing their sum, and then subtracting 6:

$$\sum_{i=1}^{12} (r_i) - 6.$$

Mathematically these sample sums are normally distributed with a mean of 0 and a standard deviation of 1:

$$\mu_\Sigma = 0.50(12) - 6 = 0$$

$$\sigma_\Sigma = 1/\sqrt{12S}\left(\sqrt{12}\right) = 1.$$

Depending on mathematical skills, the derivation processes discussed in this chapter may or may not have been easy to follow. In most instances, however, knowledge of the final result of the derivation is all that is needed.

In almost all our work, we have paid careful attention to the shapes of population distributions in assessing the applicability of statistical analyses and tests. For example, to apply a t-test when using small samples, we must establish the normality of the underlying population(s); in linear regression, we assume normality and homoscedasticity. That is, our tests thus far have been parametric, depending on the characteristics of the underlying distributions. In this chapter, we examine tests for randomness and goodness-of-fit.

We sometimes wish to determine if a sequence of values is random, with no pattern or relationship among the values, as if each had been chosen at random from some range of values, or is not random, by showing some pattern. Such investigations take the form of hypothesis tests with these hypotheses:

H_0: The values are random.

H_a: The values are not random.

The decision rule may be formulated in a variety of ways, of which we will examine four.

The Frequency Test

Suppose that a sequence of n values is to be chosen, at random and one at a time, from a list of k candidate values. At each selection of a random value to be included in the sequence, each candidate value has probability k^{-1} of being selected. The number of times a given value appears has the binomial distribution $b(n, k^{-1})$, with mean $nk^{-1} = n/k$; we would expect each of the candidate values to appear in the sequence of n selected values approximately n/k times, *if the selection at each step is in fact random.* Significant deviations of observed frequencies of the values from the predicted frequencies would suggest that the values are not random, but that some values are more likely than others.

Consider the "Chi-Square Tests of Goodness-of-Fit and Missing Data," tests based on a statistic with a chi-square distribution to compare predicted and observed frequencies.

If the null hypothesis is true and the values are truly random, then the statistic

$$\chi^2 = \sum_{i=1}^{k} \frac{(O_i - n/k)^2}{n/k} = \frac{k}{n}\sum_{i=1}^{k}\left(O_i - \frac{n}{k}\right)^2$$

where O_i is the observed frequency of each value, has a chi-square distribution with $k - 1$ degrees of freedom.

If the value of the chi-square statistic is near 0, the observed and expected frequencies must be similar, and we cannot conclude that the values are not random. If χ^2 is large, the observed and expected frequencies must differ, and the values are not random. We may use the chi-square table in the usual way to find the critical value of the chi-square statistic above which we may reject H_0 and accept H_a, concluding that the values are not random.

For example, consider this list of 50 digits from 0 to 4:

4 3 2 3 3 2 2 1 3 1 1 0 2 0 2 1 3 0

1 4 0 0 1 4 0 2 3 2 1 1 0 2 3 4 0 0

2 4 4 2 2 1 4 4 0 3 1 1 2 1

At the 5% significance level, we test the null hypothesis that these values are random against the alternative that they are not.

The digits are selected from a list of five possible values—0, 1, 2, 3, and 4—so $k = 5$, while $n = 50$. Corresponding to $k - 1 = 4$ degrees of freedom, the critical value is $\chi^2 = 9.488$.

Each expected frequency is $50 \cdot \frac{1}{5} = 10$, while the five observed frequencies are

$$O_0=10; O_1=12; O_2=12; O_3=8; O_4=8.$$

The value of the chi-square statistic is

$$\chi^2 = \frac{5}{50}\sum_{i=0}^{4}(O_i - 10)^2$$
$$= \frac{1}{10}\left[(10-10)^2 + (12-10)^2 + (12-10)^2 + (8-10)^2 + (8-10)^2\right]$$
$$= \frac{1}{10}[0 + 4 + 4 + 4 + 4]$$
$$= \frac{16}{10} = 1.6$$

This value is less than the critical value, so we do not reject the null hypothesis; we cannot conclude that these values were randomly selected.

Run this problem in Excel, by simply entering the data in cells A1 through A50 and then enter the associated formulas shown in Table 8.2.

Values	Ob Freq	Ex Freq	Diff	Diff ^ 2
0	=COUNTIF($A$1:$A$50,E7)	10	=F7–G7	=H7*H7
1	=COUNTIF($A$1:$A$50,E8)	10	=F8–G8	=H8*H8
2	=COUNTIF($A$1:$A$50,E9)	10	=F9–G9	=H9*H9
3	=COUNTIF($A$1:$A$50,E10)	10	=F10–G10	=H10*H10
4	=COUNTIF($A$1:$A$50,E11)	10	=F11–G11	=H11*H11
	chi-sq=	=(1/10)*SUM(I7:I11)		

TABLE 8.2 Excel Functions to Construct a Chi-Square Test.

The following function in Excel provides the calculation of the probability of a value occurring at the 5% significance level:

prob= =CHIDIST(1.6,4)

This generates the data found in Table 8.3.

Values	Ob Freq	Ex Freq	Diff	Diff ^ 2
0	10	10	0	0
1	12	10	2	4
2	12	10	2	4
3	8	10	–2	4
4	8	10	–2	4
	chi-sq=	1.6		

TABLE 8.3 Data File for a Goodness-of-Fit Test.

Then complete the analysis as done above. Also note that this is confirmed with the following probability, which is greater than $\alpha = 0.05$:

prob= 0.80879

The Gap Test

A *gap* in a sequence of random numbers is the number of values *between* two identical values. For example, in the sequence 6, 5, 3, 2, 3, 6, 1,…, there is a gap of length 4 between the first and second occurrence of the digit 6, and a gap of length 1 between the first and second 3. In a sequence of digits chosen randomly from a set of k candidate values, the length of a gap after a given value is a random variable X with probability function

$$f(x) = P(X = x) = \left(1 - \frac{1}{k}\right)^x \frac{1}{k}, x = 0, 1, 2, \ldots$$

That is, a gap of length x occurs after a particular value γ when the next x values are not γ, and the $(x + 1)$st value is γ. This occurs with probability

$$\left(1 - \frac{1}{k}\right)^x \frac{1}{k}.$$

If a sequence of random values contains N gaps, the expected number of gaps of length x is:

$$N * f(x) = N \left(1 - \frac{1}{k}\right)^x \frac{1}{k}.$$

To investigate the randomness of the sequence, we can compare the expected and observed numbers of gaps of each length using the chi-square statistic as in the frequency test. The associated number of degrees of freedom will be one less than the number of gaps considered.

Consider again the sequence of 50 digits from 0 to 4 tested in the previous section. There are $N = 45$ gaps in the sequence, where values were chosen from the digits 0, 1, 2, 3, and 4. Because the probability of a gap of, say, length 2 is

$$\left(1 - \frac{1}{5}\right)^2 \frac{1}{5} = \frac{16}{125} = 0.28,$$

the expected number of gaps of length 2 is

$$\frac{16}{125} * 45 = 5.76.$$

Finding the other expected gap frequencies in the same way, and counting the observed numbers of gaps, we obtain Table 8.4.

	Digits					Gap	
Length	0	1	2	3	4	Expected	Observed
0	2	3	2	1	2	10	9.0
1	1	2	1	1		5	7.2
2	1	1	2			4	5.76
3	3	1	1	1	3	9	4.61
4		2	1			3	3.69
5	1	1	2	1		5	2.94
6						0	2.36
7			1	1		2	1.89
8	1					1	1.51
9				1		1	1.21
10			1		1	2	0.97
11		1				1	0.77
12			1			1	0.62
13						0	0.49
14						0	0.40
15						0	0.32
16						0	0.25
17						0	0.20
18					1	1	0.10
						45	

TABLE 8.4 The Gap Test of Randomness.

We now compute the chi-square statistic to test the null hypothesis that the sequence of values is random against the alternative that it is not:

$$\chi^2 = \sum_{i=0}^{18} \frac{(O_i - E_i)^2}{E_i} = \frac{(10-9.0)^2}{9.0} + \frac{(5-7.2)^2}{7.2} + \ldots + \frac{(1-0.16)^2}{0.16}$$

$$= 17.114.$$

Corresponding to $19 - 1 = 18$ degrees of freedom, the value of $\chi^2_{0.05}$ is 28.9. The value of the chi-square is less than this critical value, so again we cannot reject the null hypothesis that a given sequence of values is random at the 5% significance level.

Excel contains no procedure that performs the gap test.

The Poker Test

Consider the question: "What is the probability that a five-card poker hand will contain two of a kind?" We can think of generating a random sequence of numbers as dealing from an infinite, well-shuffled deck, and apply reasoning similar to that of the card problems to develop another test of randomness.

Given a sequence that we wish to test for randomness, we break up the sequence into 5-tuples, and use a chi-square statistic to compare the expected and observed frequencies of particular types of "hands": two of a kind, three of a kind, two pair, etc. Because our "deck" is essentially infinite, finding the probabilities of the types of hands will differ from the card examples, as shown in this redo of the first example in this chapter.

Broken into 5-tuples, the sequence of numbers is this:

$$(4\ 3\ 2\ 3\ 3)(2\ 2\ 1\ 3\ 1)(1\ 0\ 2\ 0\ 2)(1\ 3\ 0\ 1\ 4)$$
$$(0\ 0\ 1\ 4\ 0)(2\ 3\ 2\ 1\ 1)(0\ 2\ 3\ 4\ 0)(0\ 2\ 4\ 4\ 2)$$
$$(2\ 1\ 4\ 4\ 0)(3\ 1\ 1\ 2\ 1)$$

When selecting digits randomly from the five alternatives (0, 1, 2, 3, and 4), the probability that an arbitrary 5-tuple will contain exactly one pair of identical values is:

$$1 * \frac{1}{5} * \frac{4}{5} * \frac{3}{5} * \frac{2}{5} * \binom{5}{2} = \frac{4!5!}{5^4 2!3!} = \frac{48}{125} = 0.384.$$

There are ten 5-tuples in the sequence, so the expected number of hands containing exactly one pair of values is 10*0.384 = 3.84. Similar calculations, and counting the occurrences of the various arrangements, yield Table 8.5.

Frequencies			
"Hand"	Probability	Expected	Observed
One Pair	0.384	3.84	3
Two Pair	0.288	2.88	4
Three of a Kind	0.192	1.92	3
Full House	0.064	0.64	0
Four of a Kind	0.032	3.32	0

TABLE 8.5 Poker Test Data File.

We find the value of the chi-square statistic in the usual way:

$$\chi^2 = \sum_{i=1}^{5} \frac{(O_i - E_i)^2}{E_i} = \frac{(3-3.84)^2}{3.84} + \frac{(4-2.88)^2}{2.88} + \frac{(3-1.92)^2}{1.92}$$

$$+ \frac{(0-0.64)^2}{0.64} + \frac{(0-0.32)^2}{0.32} = 2.036.$$

The associated number of degrees of freedom is one less than the number of "hands" considered. In this case, that value is $5 - 1 = 4$, and the value of $\chi^2_{0.05}$ is then 9.488. The observed value of the chi-square statistic is less than this critical value, so we cannot conclude at the 5% significance level that the sequence is not random.

Again, this is not a test that Excel supports.

The Runs Test

If a group of 12 men and women form a line in this order:

F   F   F   F   F   F   M   M   M   M   M   M

we would conclude immediately that their arrangement is not random, as we would if this arrangement were observed:

M   F   M   F   M   F   M   F   M   F   M   F.

Sequences composed of only two types of symbols can be tested for randomness by counting the number of *runs*, or sequences of the same symbol, in the entire sequence. For example, the sequence

0   0   1   1   1   0   1

contains a run of two, a run of three, and two runs of one for a total of four runs.

If the number of runs in a sequence is small, lack of randomness through *clustering* is indicated, as in the first example above, while *mixing*—the second example—is reflected in too many runs. Because randomness may fail for either of these two mutually exclusive reasons, and the number of runs distinguishes them, we can perform a two-sided test for randomness or either of two one-sided tests.

Two-sided test:

H_0: The sequence is random.

H_a: The sequence is not random.

One-sided tests:

lower-tail

H_0: The sequence is random.

H_a: The sequence tends to cluster.

upper-tail

H_0: The sequence is random.

H_a: The sequence tends to mix.

A sequence to be tested consists of n_1 symbols of one type and n_2 symbols of a second; the number of possible runs has a minimum of 2 and a maximum of $2n_1$ if $n_1 = n_2$ or $2n_1 + 1$ if $n_1 < n_2$. Critical values for the number of runs, found in standard statistical critical value tables, which ordinarily provide critical values of the two-sided test at the 5% significance level (or the one-sided tests at $\alpha = 2.5\%$) for n_1 and $n_2 \le 20$.

For example, this sequence of binary digits contains six runs and is composed of $n_1 = 10$ 0s and $n_2 = 12$ 1s.

$$0\ 0\ 0\ 1\ 1\ 1\ 1\ 0\ 0\ 0\ 0\ 1\ 1\ 0\ 0\ 0\ 1\ 1\ 1\ 1\ 1\ 1$$

The critical values for the two-sided test at the 5% significance level are 7 and 17; the observed number of runs is less than 7, so we conclude that the sequence is not random.

More strongly, because the number of runs is less than the critical value for the lower-tail test at the 2.5% significance level, we conclude that the sequence tends to cluster.

As N increases, the sampling distribution of U, the number of runs, approaches the normal distribution with mean $\mu_U = 1 + (2n_1 n_2)/N$ and standard deviation

$$S_U = \left[\frac{2n_1 n_2 (2n_1 n_2 - N)}{N^2(N-1)} \right]^{1/2}.$$

The critical values of U for the two-sided test at the α significance level are:

$$U^* = \mu_U \pm z_{\alpha/2} S_U.$$

Runs Above and Below a Central Value

A sequence whose entries take on more than two values may be tested for randomness with a runs test in two different ways. In the first, a dichotomy is imposed on the values in the sequence by creating a new sequence whose entries reflect whether each entry in the original sequence is above (A) or below (B) some central value in the population from which the original entries were selected, typically the mean or median. The sequence of A's and B's can be tested with the runs already given. Such a test is said to be based on *runs above and below the mean or median.*

For example, to test this sequence of 20 digits from 0 to 9, we create the corresponding sequence of symbols A and B depending on whether each entry in the sequence is above or below the population median of 4.5:

3	4	4	1	8	0	1	0	9	5	7	6	8	2	5	9	3	0	8	7
B	B	B	B	A	B	B	B	A	A	A	A	A	B	A	A	B	B	A	A

There are eight runs in the sequence of 10 A's and 10 B's; the critical values for the two-sided test at the 5% significance level are 6 and 16. The observed number of runs lies between the critical values, so we cannot conclude that the original sequence is not random.

Again, this is not a test that Excel supports.

Runs Up and Down

In dichotomizing a sequence, as was done in the previous section, information about the sequence is inevitably lost. Some of this can be preserved by considering not runs above and below a value, but directional runs, or runs up and down. Again, we create a new sequence of symbols from the original sequence, but here, every entry except the first receives a + or a – depending on whether the entry is greater or less than its predecessor. The runs test is then applied to the sequence of +s and –s. (If any entry replicates its predecessor, a 0 is entered. Zeroes are ignored in creating and counting runs.)

Applying this technique to the previous example, we generate this sequence of +s and –s:

3	4	4	1	8	0	1	0	9	5	7	6	8	2	5	9	3	0	8	7
	+	0	–	+	–	+	–	+	–	+	–	+	–	+	+	–	–	+	–	–

In the sequence of nine +s and nine –s, we find $U = 16$ runs. Using a standard table for the U statistic the critical values for the two-sided test at $\alpha = 5\%$ are 5 and 15. U is not between these critical values, so we can conclude from this test that the original sequence of values is not random. In particular, because U is greater than the upper critical value, we can conclude that the sequence is not random because the direction changes too often.

This procedure is not performed by Excel.

The Kolmogorov Goodness-of-Fit Test

One of the many applications of the chi-square distribution is the goodness-of-fit test, which is used to investigate whether a set of values might have come from a specified distribution. Another technique, developed by the Russian mathematician, A.N. Kolmogorov, tests for goodness-of-fit by comparing the empirical *cumulative distribution function* (CDF) with the hypothesized CDF, using the largest absolute difference of the two functions as the test statistic, usually called D. As with the chi-square test, the null hypothesis is that the values come from the proposed distribution, while the alternative is that they come from some other distribution.

The value of the cumulative *empirical* distribution at any data point x is

$$S(x) = \frac{N(x)}{N},$$

where $N(x)$ is the number of values less than or equal to x, and N is the total number of data values.

Two important observations should be made about the Kolmogorov test. First, the distribution of the test statistic D is not dependent on the nature of the underlying population distribution. Second, while the chi-square test of goodness-of-fit requires grouping possible values of a continuous distribution into classes, no such grouping is required here. Thus, it can be argued that the Kolmogorov test is better suited to situations involving continuous distributions.

The Kolmogorov-Smirnov Two-Sample Test

In a procedure very much like the Kolmogorov test, we can examine whether two independent samples might have come from identical distributions. This test was developed by another Russian, N.V. Smirnov, and generally carries the names of both men.

Given two independent samples, we construct the two empirical cumulative distribution functions $S_1(x)$ and $S_2(x)$, and use the statistic $D = $ maximum $|S_1(x) - S_2(x)|$ to test the hypotheses:

H_0: The samples come from identical populations.

H_a: The populations are not identical.

The calculation of D is more straightforward than in the Kolmogorov one-sample test.

Hubert's Γ Statistic

Hubert's Γ statistic is useful for assessing fit between data and expectant structures. Hubert & Schultz[4] provide an overview of the statistic complete with examples.

Let $A = [x(i,j)]$ and $B = [y(i,j)]$ care two $n \times n$ proximity matrices on the same set of n objects. Let

$$y(i,j) = \begin{cases} 0, \textit{if object i and have the same category label} \\ 1, \textit{if not} \end{cases}.$$

The Hubert Γ statistic is the point serial correlation between the two proximity matrices in normalized form:

$$\Gamma = \sum_{i=1}^{n-1} \sum_{j=i+1}^{n} X(i,j)Y(i,j).$$

Let m_x and m_y be the sample means and s_x and s_y be the sample standard deviations of the values in the matrices A and B. Then

$$\Gamma = \left\{ \left(\frac{1}{M} \right) \sum_{i=1}^{n-1} \sum_{j=i+1}^{n} \left[X(i,j) - m_x \right] \left[Y(i,j) - m_y \right] \right\} \Big/ s_x s_y$$

where $M = n(n-1)/2$ is the number of entries and:

$$m_x = (1/M) \sum \sum X(i,j) \qquad m_y = (1/M) \sum \sum Y(i,j)$$

$$s_x^2 = (1/M) \sum \sum X(i,j)^2 - m_x^2 \qquad s_y^2 = (1/M) \sum \sum Y(i,j)^2 - m_x^2$$

all sums are over the set $\{(i,j): 1 \le i \le n - 1; i + 1 \le j \le n$.

---

[4] Hubert, L. J. and Schultz, J. (1976). Quadratic assignment as a general data-analysis strategy. *British Journal of Mathematical and Statistical Psychology* 29, 190-241.

8.4 INDICES OF CLUSTER VALIDITY

An *external validity index* is a measure addressing the situation where the classification for the set of objects already exists and the measurement taken is the degree to which the clustering method generates a clustering matching the original classification scheme. For example, consider the problem of determining whether or not a hierarchical clustering obtained in a cluster analysis for a specific data set matches an expected hierarchical clustering. Because this requires that in a hypothesis test that the null distribution is dependent on a number of factors (number of objects, type of population, type of hierarchical clustering method employed, etc.), then the expected hierarchy usually is not accessible. As a result, the need to acquire external indices is rare. Therefore this topic is not discussed at this time and the interested reader should do a research study on this topic.

Another manner in which to validate results is to measure the degree to which a partition obtained from a clustering method is justified by the given proximity matrix. Using only the data, an *internal validity index* measures the fitness of the clustering structure with respect to the data. A determination of whether or not a hierarchical clustering matches an expected hierarchical clustering needs to be computed. Both hierarchies can be expressed as proximity matrices. The expected hierarchy is a dissimilarity matrix in which the entry in row i and column j is k if the objects j and i are expected to be first in the same cluster at level k of the hierarchy, which is available in the hierarchy's dendrogram. The captured hierarchical clustering can be represented by *cophenetic proximity matrix*.

Let $\{C_1, C_2, \ldots, C_n\}$ be a hierarchical clustering where C_n contains $n - m$ clusters and L, whose value is in the dendrogram, is a level function which is set equal to the proximity at which each clustering is formed.

$$L(m) = \min \{d(x_i, x_j): C_m \text{ is defined.}\}$$

The cophenetic proximity measure

$$d_C(i,j) = L(k_{ij})$$

where

$$k_{ij} = \min \{m : (x_i, x_j): C_{mq} \text{ for some } q.$$

The cophenetic matrix contains cell values of $d_c(i,j)$. The nearest neighbor and farthest neighbor generate the same dendrogram when the clusterings are derived from a cophenetic proximity matrix.

A product-moment correlation coefficient, called the cophenetic correlation coefficient, CC, can be employed to determine the correlation between the cophenetic and the hierarchy's proximity matrix. This is a symmetrical matrix, therefore, only the upper triangle of $m = n(n - 1)/2$ needs to be posted for use in the computation of the cophenetic correlation coefficient.

$$CC = \frac{(1/M)\sum d(i,j)d_C(i,j) - (m_D m_C)}{\left[(1/M)\sum d(i,j)^2 - m_D\right]^{1/2}\left[(1/M)\sum d_C(i,j)^2 - m_C\right]^{1/2}}$$

where $m_D = \left(\frac{1}{M}\right)\sum d(i,j)$, $m_D\left(\frac{1}{M}\right)\sum d_C(i,j)$, and all sums are over the set $\{(i,j): 1 \le i < j \le n\}$.

The value of the CC is between –1 and 1, with 1 representing a perfect positive match.

A means for determining the internal validity is to run a Monte Carlo study as outlined by Jain & Dubes,[5] as is the material on the cophenetic matrix and correlation coefficient. The following algorithm is from page 168.

ALGORITHM FOR BASELINE DISTRIBUTION OF 7 UNDER RANDOM GRAPH HYPOTHESIS

Step 1. For fixed n (number of objects) form a dissimilarity matrix under the random graph hypothesis; that is, fill in the $n(n - 1)$ 12 entries with a randomly chosen permutation of the integers from 1 to $n(n - 1)/2$.

Step 2. Cluster the n objects by a clustering method, such as the single-link or complete-link method.

Step 3. Form the cophenetic matrices for the dendrogram resulting from the clustering method.

Step 4. Compute y between the dissimilarity and cophenetic matrices.

---

[5] Jain, A. K. & Dubes, B. C. (1988). *Algorithms for clustering data.* Englewood Cliffs, NJ: Prentice Hall.

Repeat Steps 1 to 4 on a Monte Carlo basis to create a baseline distribution for $-y$ specific to the clustering method and value of n. Hubert[6] used 1,000 trials for each value of n.

Often in cluster analysis the investigator needs to decide which is the "best" clustering among a set of captured clusterings. How well any of the clusterings fits the data is not being considered, but, rather, consider which member of the collection of clusterings fits the data the best. A *relative validity index* enables an investigator to answer this question.

Rand[7] was one of the first to develop a measure of similarity between two clusterings (Y & Y') of the same data. He based his statistic on three assumptions.

First, clustering is discrete in the sense that every point is unequivocably assigned to a specific cluster. Second, clusters are defined just as much by those points which they do not contain as by those points which they do contain. Third, all points are of equal importance in the determination of clusterings (p. 847).

From these assumptions, it follows that a basic unit of comparison between two clusterings is how pairs of points are placed. There are four possibilities (types): (a) the pair is placed in the same cluster for both methods; (b) the pair is placed in the same cluster by one method (Y) and different clusters by the second method (Y1); (c) the pair is placed in the same cluster by the second method (Y1) and different clusters by the first method (Y); and (d) the pair is placed in different clusters by both methods. Cases (a) and (d) above represent similarity between the clusterings and cases (b) and (c) represent dissimilarity. Rand's measure of similarity, c(Y, Y'), can be defined as the number of similar assignments of point-pairs normalized by the total number of point-pairs. More specifically, given N points $(X_1, X_2, \ldots, X_N)$ and two clusterings of them $Y = \{Y_1, \ldots, Y_{K1}\}$ and Y1 $= \{Y_1, \ldots, Y_{K2}\}$, by definition

$$c(Y, Y') = \sum_{i<j}^{N} Y_{ij} \Big/ \binom{N}{2}, \quad C_1$$

---

[6] Hubert, J. (1974). Approximate evaluation techniques for the single-link and complete-link hierarchical clustering structures. *Journal of the American Statistical Association, 69,* 698-704.

[7] Rand, W. M. (1971). Objective criteria for the evaluation of clustering methods. *Journal of the American Statistical Association, 66,* 845-850.

where

Y_{ij} = 1 if there exist K and K' such that both x_i and x_j are in both Y_K and $Y_{K'}$.

= 1 if there exist K and K' such that x_i is in both Y_K and Y'_K, while x_j is in neither Y_K or Y'_K

= 0 otherwise, and $\begin{vmatrix} a \\ b \end{vmatrix}$ represents the usual binomial coefficient.

Unfortunately, the Rand index has been shown to have some undesirable properties. For example, the index approaches its upper bound of 1 as the number of clusters increases without limit (Milligan, Soon, & Sokol).[8] The index also fails to take into account chance agreement (Morey & Agresti).[9] Using a Monte Carlo approach (Milligan and Schilling),[10] and (Milligan, Soon, and Sokol)[2,8] compared four external criterion measures under a number of different conditions. The four methods were the Rand,[1] the adjusted Rand (Morey and Agresti),[3] the Jaccard and the Fowlkes and Mallows[11] statistic, reported in Milligan & Shilling.[4] Based on the four possibilities mentioned above (a, b, c, and d), the following four equalities will be used to define the four measures:

$$a = \left| \frac{\sum_i \sum_j N_{ij}^2}{2} \right| - \left(\frac{N..}{2} \right),$$

$$b = \left| \frac{\sum_i N_i^2}{2} \right| - \left| \frac{\sum_i \sum_j N_{ij}^2}{2} \right|,$$

$$c = \left| \frac{\sum_j N_j^2}{2} \right| - \left| \frac{\sum_i \sum_j N_{ij}^2}{2} \right|,$$

[8] Milligan, G. W., Soon, T., & Sokol, L. (1983). The effect of cluster size, dimensionality and the number of clusters on recovery of the cluster structure. *IEEE Transactions on Pattern Analysis and Machine Intelligence, 5*, 40-47.

[9] Morey, L., & Agresti, A. (1984). The measurement of classification agreement: An adjustment to the Rand statistic for chance agreement. *Educational and Psychological Measurement, 44*, 33-37.

[10] Milligan, G. W., & Schilling, D. A. (1985). Asymptotic and finite sample of error perturbation on fifteen clustering algorithms. *Psychometrida, 45*(3), 325-342.

[11] Fowlkes, E. F., & Mallows, C. L. (1983). Rejoinder. *Journal of the American Statistical Association, 78*, 584.

$$d = \left| \left. \sum_i \sum_j N_{ij}^2 \middle/ 2 \right| + \left(N.. \middle/ 2 \right) - \left| \left. \sum_j N_i^2 \middle/ 2 \right| - \left| \left. \sum_i N_j^2 \middle/ 2 \right| \right. \right.,$$

where N_{ij} is the number of points in cluster i as produced by the first algorithm which is also in cluster j of the second algorithm (or the true criterion solution). Also N_i, N_j, and $N..$ represent the marginal and grand totals. The four statistics can now be defined as follows:

Rand: $[a + d]/[a + b + c + d]$

Adjusted Rand: $[a + d - Nc]/[a + b + c + d - Nc]$

Jaccard: $[a]/[a + b + c]$

Fowlkes and Mallows: $[a]/[(a + b)(a + c)]^{1/2}$

where Nc is defined as follows:

$$\left| \left. \sum_i \sum_j N_i.^2 N_{.j}^2 \middle/ N..^2 \right| + \left[N..(N.. - 1) \middle/ 2 \right] - \left\{ \left. \sum_i N_i^2 \middle/ 2 \right\} - \left\{ \left. \sum_j N_j^2 \middle/ 2 \right\} \right. \right. . \quad C_2 \right.$$

The results of both studies indicated that the Jaccard and the adjusted Rand performed superior to the other measures. Milligan and Schilling[4] concluded by saying that "When selecting indices for future research, it would seem reasonable to use the adjusted Rand and Jaccard measures. The enhanced variability and sensitivity of the two statistics seems to provide the best characteristics for the measurement of cluster recovery" (p. 108).

Hubert and Arabic[12] derive a different correction for chance for the Rand statistic than the adjusted Rand (Morey & Agresti).[3] They state that "Probably the most obvious (null) model for randomness assumes that the $R \times C$ contingency table is constructed from the generalized hypergeometric distributions, i.e., the U and V [Y and Y1 in our case] partitions are picked at random, subject to having the original number of classes and objects in each. Under the hypergeometric assumption, we can show:

$$E\left(\sum_{i,j} \left\langle \begin{matrix} n_{ij} \\ 2 \end{matrix} \right\rangle \right) = \sum_i \left\langle \begin{matrix} n_i. \\ 2 \end{matrix} \right\rangle \sum_j \left\langle \begin{matrix} n_{.j} \\ 2 \end{matrix} \right\rangle \middle/ \left\langle \begin{matrix} n.. \\ 2 \end{matrix} \right\rangle \quad C_3$$

---

[12] Hubert, L., & Arabic, P. (1985). Comparing Partitions. *Journal of Classification, 2*, 193-218.

where $\left\langle\genfrac{}{}{0pt}{}{a}{b}\right\rangle$ represents the usual binomial coefficient. In equation C3 and subsequent equations, the a corresponds to the n_{ij}s and its variations and the b corresponds to the 2s. In this notation n_{ij} is equivalent to N_{ij} of the previous equations. The same is true for the marginals $n_{i.}$ and $n_{.j}$. $N_{..}$ is equivalent to n, the grand total (n). Equation C3 gives the expected number of object pairs of type (a), i.e., pairs in which the objects are placed in the same cluster in Y and in the same cluster in Y′. The value to the right of the equal sign is the number of distinct pairs that can be constructed within rows, times the number of distinct pairs that can be formed from columns, divided by the total number of pairs.

They go on to show that the Rand measure has expectation:

$$E\left(A\Big/\left\langle\genfrac{}{}{0pt}{}{n}{2}\right\rangle\right)=+2\sum_i\left\langle\genfrac{}{}{0pt}{}{n_i}{2}\right\rangle\Big/\sum_j\left\langle\genfrac{}{}{0pt}{}{n_j}{2}\right\rangle\Big/\left\langle\genfrac{}{}{0pt}{}{n}{2}\right\rangle^2-\left|\sum_i\left\langle\genfrac{}{}{0pt}{}{n_i}{2}\right\rangle+\sum_j\left\langle\genfrac{}{}{0pt}{}{n_j}{2}\right\rangle\right|\Big/\left\langle\genfrac{}{}{0pt}{}{n}{2}\right\rangle, \quad (C_4)$$

where A is equal to (a) + (d) from the four types listed above, i.e., the total number of agreements. By using the general form of an index corrected for chance, which is given by:

(Index – Expected Index)/(Maximum Index – Expected Index), (C_5)

and is bounded above by 1 and takes on the value of 0 when the index equals its expected value, the corrected Rand index would have the form (assuming a maximum Rand index of 1):

$$\sum_{ij}\left\langle\genfrac{}{}{0pt}{}{n_{ij}}{2}\right\rangle=\sum_i\left\langle\genfrac{}{}{0pt}{}{n_i}{2}\right\rangle\sum_j\left\langle\genfrac{}{}{0pt}{}{n_j}{2}\right\rangle\Big/\left\langle\genfrac{}{}{0pt}{}{n}{2}\right\rangle\Big/\frac{1}{2}$$

$$\left\{\left|\sum_i\left\langle\genfrac{}{}{0pt}{}{n_i}{2}\right\rangle+\sum_j\left\langle\genfrac{}{}{0pt}{}{n_j}{2}\right\rangle\right|-\left|\sum_i\left\langle\genfrac{}{}{0pt}{}{n_i}{2}\right\rangle\sum_j\left\langle\genfrac{}{}{0pt}{}{n_j}{2}\right\rangle\Big/\left\langle\genfrac{}{}{0pt}{}{n}{2}\right\rangle\right|\right\} \quad (C_6)$$

Hubert and Arabic conclude the section by saying "as defined, the Morey-Agresti correction inappropriately assumes that the expectation of a squared random variable is the square of the expectation. Specifically, Morey and Agresti assert that

$$E\left(\sum_{ij}n_{ij}^2\right)=\sum_{ij}n_i^2n_j^2/n^2, (C_7),$$

whereas our equation C3 could be rewritten to show

$$E\left(\sum_{ij}n_{ij}^2\right)=\left|\sum_{ij}n_i^2n_j^2/(n(n-1)+n^2/(n-1)\right|-\left\{\sum_i n_i^2+\sum_j n_j^2\right\}/(n-1). \quad (C_8)$$

In general, equation C8 is larger than equation C7 and the positive difference of

$$\left[\frac{1}{n^2(n-1)}\right]\left\{n^2 - \sum_i n_i^2\right\} \Big/ \left\{n^2 - \sum_j n_j^2\right\}, \quad (C_9)$$

is not necessarily small, depending on the sizes of the object sets and associated partitions being compared" (p. 200). Hubert and Arabic went on to give a brief example of the differences between the three different Rand indexes.

8.5 SUMMARY

- Determining the validity of a clustering is basically a statistical problem.

- Seldom does the investigator have full knowledge for the required baseline distribution, for determining the validity of a clustering or when comparing clustering structures, and therefore, the baseline distribution must be implemented by Monte Carlo methods.

- An external validity index is used to determine a clustering structure against expectant information.

- An internal validity index uses only the proximity matrix for the clustering and information from the cluster analysis to determine the validity of the clustering.

- A relative validity index compares two clustering methods.

- The validation of clustering structures is vital to the formalization and strengthening of the results of cluster analysis studies. Rather than simply being an art, the validation is a step toward enabling cluster analysis to become a science.

8.6 EXERCISES

1. We will take a sample of size 80 from a population whose standard deviation we know to be 56 and test these hypotheses:

$$H_0: \mu \leq 300$$
$$H_a: \mu > 300$$

   a. At the 5% significance level, find the critical value and state the decision rule.

b.   For these possible values of μ, find the probability of a Type II error: 305, 300, 315, 320.

c.   Use the values found in part b to sketch the operating characteristic and power curves of the test.

d.   Find the critical value corresponding to $\alpha = 1\%$, restate the decision rule, and find β if $\mu = 315$. What generalizations does this suggest about α and β?

e.   In the original test, $\overline{X} = 312$. What conclusion is reached? What type of error might have been made?

f.   Perform the test in part e again by comparing the z statistic to the critical normal deviate, then by comparing the probability that $\overline{X}$ would be at least as large as its observed value with α.

2. Show that for any upper-tail test of the population mean, the value of β if $\mu = \overline{X}^*$ is 0.5000.

3. What factors influence the power of a test of hypotheses?

4. Use the following data from an American Cities database to test the claim that construction in U.S. cities was up by more than 5% at the 5% significance level:

Variablex7		Change in Construction Activity			
Mean	5.523	Std err	1.081	Std dev	9.301
Variance	86.517	Kurtosis	1.947	Skewness	.924
Minimum	−15.400	Maximum	40.600	Sum	408.700
C.V.Pct	168.414	.95 C.I.	3.368	To	7.678
Valid cases	74	Missing cases	0		

TABLE 8.6 Descriptive Statistics for Change in Construction Activity.

a.   Formally state the hypotheses and determine the decision rule.

b.   If construction activity has in fact increased by 5.5%, find the probability that we will fail to reject H_0. Illustrate by drawing the distribution of $\overline{X}$ if $\mu = 5.5\%$ and indicate the area β.

c.   Sketch the OC and power curves for this test.

d.   Use the decision rule to come to a conclusion. What type of error might have been made?

5. Show that for any lower-tail test of hypotheses for μ, β is 0.50000 if $\mu = \overline{X}^*$.

6. Find $z_{\alpha/2}$ for $\alpha = 1\%, 2\%, 5\%$, and 10%.

7. A table of random numbers contains 15,050 digits, which should have been chosen at random from the digits 0 through 9. The following table gives the frequencies of each digit in the table. At the 5% significance level, should it be concluded from this data that the table is not random?

Digit	0	1	2	3	4	5	6	7	8	9
Frequency	1493	1491	1461	1552	1494	1454	1613	1491	1482	1519

8. Use the constant multiplier technique with $K = 6787$ and $X_0 = 4129$ to obtain a sequence of three four-digit random numbers.

9. Let a sequence of random numbers $(R_1, R_2, R_3, R_4, R_5)$ be 0.45, 0.37, 0.89, 0.11, and 0.66. Extend the sequence through R_{10} using the additive congruential method, where $m = 100$.

10. Determine whether the historical linear congruential generators shown below can achieve a maximum period. Conduct an Internet search. Also, state restrictions on X_0 to obtain this period.

    a.   In SIMSCRIPT for CDC; $a = 2{,}814{,}749{,}767{,}109$; $c = 59{,}482{,}661{,}568{,}369$; $m = 2^{48}$

    b.   IBM 360; $a = 69{,}019$; $c = 0$; $m = 2^{33}$

    c. $a = 6507$; $c = 0$; $m = 1024$

11. Use the mixed congruential method to generate a sequence of three two-digit random numbers with $X_0 = 37$, $a = 7$, $c = 29$, and $m = 100$.

12. Use the mixed congruential method to generate a sequence of three two-digit random integers between 0 and 24 with $X_0 = 13$, $a = 9$, and $c = 35$.

13. Consider the multiplicative congruential generator under the following circumstances:

    a. $a = 11$, $m = 16$, $X_0 = 7$

    b. $a = 11$, $m = 16$, $X_0 = 8$

    c. $a = 7$, $m = 16$, $X_0 = 7$

    d. $a = 7$, $m = 16$, $X_0 = 8$

Generate enough values in each case to complete a cycle. What inferences can be drawn? Is maximum period achieved?

14. Generate five random observations from a *uniform distribution* between −10 and +40.

15. Suppose that random observations are needed from the *triangular distribution* whose probability density function is

$$f(x) = \begin{bmatrix} 2x, \text{ if } 0 < x < 1 \\ 0, \text{ otherwise} \end{bmatrix}$$

a. Derive an expression for each random observation as a function of the random decimal number *r*.

b. Generate five random observations.

16. Apply the randomness test of runs above and below the median to this sequence of 15 values, using the 5% significance level:

22, 2, 4, 12, 11, 15, 28, −5, 8, 4, −1, −10, −2, 25, 7

17. A programmer has developed a batch of 14 programs, some of which contain bugs. This sequence shows which contain bugs and which do not:

B B NB NB NB B NB NB NB NB B NB B B

At the 5% significance level, are the programs randomly arranged with respect to containing bugs? (Use the runs test.)

18. Apply the test of runs up and down to this sequence, at the 5% significance level:

2 4 8 2 9 4 0 7 2 5

7 5 9 0 7 7 5 9 6 0

19. Apply the gap test of randomness to this sequence, chosen from the digits, 0, 1, 2, and 3. Use the 10% significance level.

1 0 1 1 1 1 2 0 0 3

3 1 3 1 1 0 0 0 0 2

20. This sequence was selected from the digits 0 through 9. Apply the poker test of randomness to it at the 5% significance level.

1 0 0 9 7
7 6 5 2 0
3 4 6 7 3
8 0 9 5 9
3 9 2 9 2
3 2 5 3 3
1 3 5 8 6
5 4 8 7 6
0 9 1 1 7
7 4 9 4 5

TABLE 8.7 Hypothetical Poker Test Data.

21. A sequence of 60 values is chosen from a distribution purported to be the discrete uniform on $\{0, 1,..., 9\}$. These frequencies are observed:

Digit	0	1	2	3	4	5	6	7	
Frequency	9	10	4	2	3	5	1	7	11

    a.   Test the assertion of randomness from $\{0, 1,..., 9\}$ with the chi-square goodness-of-fit test at the 5% significance level.

    b.   Perform the test in part a using the Kolmogorov test at the 5% significance level. Is this test appropriate here?

22. Employees at two computer centers are categorized in five job classifications as shown in Table 8.8.

	Systems Engineers	Systems Analysts	Systems Programmers	Applications Programmers	Operators
Center A	2	0	12	14	2
Center B	4	9	18	22	7

TABLE 8.8 Computer Store Center Personnel Data.

Using the Kolmogorov-Smirnov two-sample test, can it be concluded at the 5% significance level that jobs are distributed differently in the two centers?

23. Table 8.9 lists the thousands of kilometers of railroad track available in North and Central America and in South America. Use the Mann-Whitney test at the 5% significance level to test whether the mean length of track is the same in the two regions.

Group 1: North and Central America	1,000 km	Group 2: South America	11,000 km
Canada	70.1	Argentina	40.2
Costa Rica	0.6	Bolivia	3.4
Cuba	14.5	Brazil	23.9
El Salvador	0.6	Chile	9.0
Guatemala	0.8	Colombia	3.4
Honduras	1.1	Ecuador	1.1
Mexico	19.2	Paraguay	0.5
Nicaragua	0.3	Peru	2.1
Panama	0.7	Uruguay	3.0
United States	332.8	Venezuela	2.0

TABLE 8.9 North Versus South Railway.

24. Perform the test of problem 4 using a t-test for independent samples. Which test is more appropriate here? Why?

25. A computer center director and a senior programmer each rated a group of 16 junior programmers on their software documentation skills. The ratings are provided in Table 8.10.

Junior Programmers																
	1	2	3	4	5	6	7	8	9	10	11	12	13	14	15	16
Director	4	4	5	5	3	2	5	3	1	5	5	5	4	5	5	5
Senior Programmer	2	3	3	3	3	3	3	3	2	3	2	2	5	2	5	3

TABLE 8.10 Ratings of Junior Programmers Software Documentation Skills.

Use the Wilcoxon matched-pairs signed-rank test at the 5% significance level to test for a difference in the average rating by these two individuals.

26. a. Perform the test in exercise 6 using the t-test for matched-pairs.

b. What assumption does the Wilcoxon test make about the underlying distributions?

c. In the Wilcoxon test, pairs that tie are discarded. What is the problem with this procedure? Suggest an alternate procedure.

d. Is the Wilcoxon test or the t-test more appropriate to the data in problem 6?

27. Twenty-three C# programmers are randomly assigned to three study groups. The groups are given a programming assignment. Group 1 was told to use a nonstructured, bottom-up approach, Group 2 was to use a structured, top-down approach, and Group 3 was given no specific instructions. The times in hours required by each individual to complete the project are shown in Table 8.11.

Times in Hours							
Group 1	9	8	8	6	5	4	4	4
Group 2	6	6	5	4	4	3	2	2
Group 3	8	8	7	7	7	6	5	

TABLE 8.11 Group Project Completion Times.

Use the Kruskal-Wallis test at the 5% significance level to decide if the average time varied over the three groups.

28. a. Perform the test in problem 27 using one-way analysis of variance.

b. What assumptions required by analysis of variance need not be satisfied when using the Kruskal-Wallis test?

c. How is the Kruskal-Wallis test affected by a large number of ties?

29. Sixteen software packages were ranked by two computer centers as shown in Table 8.12:

Package																
	A	**B**	**C**	**D**	**E**	**F**	**G**	**H**	**I**	**J**	**K**	**L**	**M**	**N**	**O**	**P**
Center 1	1	2	3	4	5	6	7	8	9	10	11	12	13	14	15	16
Center 2	2	1	8	6	4	3	11	14	9	5	12	7	15	16	10	13

TABLE 8.12 Computer Center Software Package Ratings.

a. Calculate the Spearman rank correlation coefficient for this data. At $a = 5\%$, is it significant?

b. Calculate the Pearson product-moment correlation coefficient for the given values.

30. RiskAMP is a full-featured Monte Carlo simulation engine for Excel. With this add-in, you can add Risk Analysis to your spreadsheet models quickly and easily. Visit the following Website to download RiskAMP: *http://www.riskamp.com/home?gclid=CJTx6YDXrp0CFYNX2god GkRAyw.*

31. Using the add-in from problem 30, calculate the time it takes to commute from home to the office. Suppose your commute to work consists of the following:

- Drive 2 miles on a highway, with 90% probability you will be able to average 65 MPH the whole way, but with a 10% probability that a traffic jam will result in average speed of 20 MPH.

- Come to an intersection with a traffic light that is red for 90 seconds, then green for 30 seconds.

- Travel 2 more miles on a surface street, averaging 30 MPH with a standard deviation of 10 MPH.

   You want to know how much time to allow for your commute in order to have a 75% probability of arriving at work on time. Additionally, when you have an important meeting, you want to know how early you need to leave the house in order to have a 99.5% probability of arriving on time.

32. Look up the following articles and write a literature review for each article:

- Hertz[13] article discusses the application of Monte Carlo methods to corporate finance.

- The paper by Boyle[14] pioneered the use of simulation in derivative valuation.

---

[13] Hertz, D. B. (January-February, (1964). Risk analysis in capital investment. *Harvard Business Review*, 95-106.

[14] Boyle, P. (May, 1977). Options: a Monte Carlo approach. *Journal of Financial Economics*, 323-338.

33. Using the Neymann-Scott cluster generation program listing in Appendix C design and implement a Monte Carlo study to evaluate the effect of neighborhood shape has on a specific type of hierarchical clustering.

34. Perform a Monte Carlo study of the K-means clustering method on different sizes of clusters.

35. Perform a Monte Carlo study of the fuzzy c-means clustering method on different sizes of clusters.

36. Perform a Monte Carlo study of the K-means clustering method on different density functions.

37. Perform a Monte Carlo study of the fuzzy c-means clustering method on different sizes of clusters.

38. Perform a Monte Carlo study of the K-means clustering method where outliers are present.

39. Given the following ordinal proximity matrix:

	2	3	4	5
1	4	6	1	5
2		8	9	3
3			7	10
4				2

TABLE 8.13 Ordinal Proximity Matrix.

First generate the single-link and complete-link dendrograms. Use these dendrograms to compute the cophenetic correlation coefficient value and interpret the result.

40. Use the Neymann-Scott cluster generation program listing in Appendix C to generate an expectant clustering of objects. Generate a hierarchical clustering using the Ward method on the same data set. Use Hubert's Γ statistic for assessing the fit between the data set and the expectant structure.

41. Use the Neymann-Scott cluster generation program listing in Appendix C to generate an expectant clustering of objects. Generate two distinct hierarchical clusterings on the same data set. Use Rand and Corrected Rand statistics for comparing the two hierarchies.

42. Use the Neymann-Scott cluster generation program listing in Appendix C to generate an expectant clustering of objects. Generate both a K-means and a fuzzy c-means clusterings on the same data set. Use Rand and Corrected Rand statistics for comparing the two clusterings.

CHAPTER 9

CLUSTERING CATEGORICAL DATA

In This Chapter

9.1 INTRODUCTION

In the previous chapters, the clustering methods discussed were built on a foundation of a similarity measure or a distance metric. This foundation necessitates that the data set be at least ordinal data. When dealing primarily with numerical data, such as in statistical studies, many metrics are usually available. Databases, on the other hand, often contain categorical data. Due to the lack of an order being present, a distance measure cannot be defined for categorical data.

Reconsider the weather data set in Table 9.1.

Day	Outlook	Temperature	Humidity	Wind	Play
D1	Sunny	Hot	High	Weak	No
D2	Sunny	Hot	High	Strong	No
D3	Overcast	Hot	High	Weak	Yes
D4	Rain	Mild	High	Weak	Yes
D5	Rain	Cool	Normal	Weak	Yes
D6	Rain	Cool	Normal	Strong	No
D7	Overcast	Cool	Normal	Strong	Yes
D8	Sunny	Mild	High	Weak	No
D9	Sunny	Cool	Normal	Weak	Yes
D10	Rain	Mild	Normal	Weak	Yes
D11	Sunny	Mild	Normal	Strong	Yes
D12	Overcast	Mild	High	Strong	Yes
D13	Overcast	Hot	Normal	Weak	Yes
D14	Rain	Mild	High	Strong	No

TABLE 9.1 Weather Data Set.

How can the distance between a "rain, mild, high" tuple and another tuple that is "sunny, cool, normal" be defined? The following major categorical data clustering algorithms that have resolved this question: ROCK,[1] STIRR,[2] CACTUS,[3] and CLICK.[4]

A common feature of these algorithms is that they model the similarity of categorical attributes. All of these methods operationally define the tuples as similar if the items with which they simultaneously occur are large.

[1] Guha, S., Rajeev, R., & Shim, K. (March, 1999). ROCK: A robust clustering method for categorical attributes. Proceedings of the IEEE International Conference on Data Engineering, Sydney, 512-521.

[2] Gibson, D., Kleinberg, J., & Raghavan. (1998). Clustering categorical data: an approach based on dynamical systems. In Proceedings of the 24th VLDB Conference, New York, USA, 311-322.

[3] Ganti, V., Gehrke, J., & Ramakrishnan, P. (1999). CACTUS: Clustering categorical data using summaries. In Proceedings of ACM SIGKDD International Conference on Knowledge Discovery and Data Mining, San Diego, CA, USA, 73-83.

[4] Peters, M. and Zaki, M. J. (2004). CLICK: Clustering Categorical Data using K-partite Maximal Cliques, In IEEE International Conference on Data Engineering. IEEE.

Links and neighbors are used for the items in ROCK, where two items have a link if they have a common neighbor. A weighted node with the weights of simultaneous items being propagated, in STIRR, has two items sharing common simultaneous occurring items exhibiting similar magnitude of weights. Both CACTUS and CLICK also employ occurrences as the basis for operationally defining similarity.

Huge data sets, found in databases, are worthless unless one can extract useful information and understand the hidden meaning in the data. Therefore, there exists a need to extract information that can support business decisions. More important is to understand the rules that have generated those data. There can be hidden patterns and trends that, if uncovered, can be used in many different areas.

ROCK is a hierarchical algorithm that can be applied to categorical data. It relies on a distance metric that can be changed to accommodate any new data and scalable for accommodating with very large databases.

STIRR is an approach based on an iterative method for assigning and propagating weights on the categorical values in a table that can be studied analytically in terms of certain types of nonlinear dynamical systems. The algorithm represents each attribute value as a weighted vertex in a graph. Starting the initial conditions, the system is iterated until a "fixed point" is reached. When the fixed point is reached, the weights in one or more of the "basins" isolate two groups of attribute values on each attribute.

9.2 ROCK

ROCK (RObust Clustering using linKs) is an agglomerative hierarchical clustering method based upon the concept of links. The number of neighbors that two tuples have, in the data set, is set equal to the number of links between the two tuples. Initially, the algorithm determines the number of links between every pair of tuples in the data set. Then an agglomerative hierarchical clustering process, starting with singleton clusters of tuples, is performed on the data set using a goodness measure for merging tuples. The termination for the process is either to stop at a predetermined number of clusters of tuples or when no links remain between the clusters of tuples. ROCK partitions the whole data set based on a sample of tuples drawn from the whole data set.

The ROCK algorithm will be illustrated for the weather data set, which has the following categorical variables each with a listing of their value sets:

Outlook: (sunny, overcast, rainy)

Windy: (true, false)

Play: (yes, no)

Temperature and Humidity have real domains. These two attributes have been transformed into new categories by partitioning Temperature (60 to 64, 65 to 74, 75 to 89, 90 to 100) as "cold, mild, warm, hot" and Humidity (65 to 74, 75 to 89, 90 to 100) as "normal, soggy, saturated." This generates the revised table illustrated in Table 9.2.

Day	Outlook	Temperature	Humidity	Windy	Play
D1	Sunny	Warm	Soggy	False	No
D2	Sunny	Warm	Saturated	True	No
D3	Overcast	Warm	Soggy	False	Yes
D4	Rainy	Mild	Saturated	False	Yes
D5	Rainy	Mild	Soggy	False	Yes
D6	Rainy	Cold	Normal	True	No
D7	Overcast	Cold	Normal	True	Yes
D8	Sunny	Mild	Saturated	False	No
D9	Sunny	Mild	Normal	False	Yes
D10	Rainy	Warm	Soggy	False	Yes
D11	Sunny	Warm	Soggy	True	Yes
D12	Sunny	Warm	Soggy	True	Yes
D13	Overcast	Mild	Saturated	True	Yes
D14	Overcast	Warm	Saturated	False	Yes
D15	Rainy	Mild	Saturated	True	No

TABLE 9.2 Revised Weather Data Set Totally Categorical.

The similarity between two tuples in the above table, t_i and t_j can be defined as:

$$\text{sim}(t_i, t_j) = \frac{|t_i \cap t_j|}{|t_i \cup t_j|},$$

where $\left|t_i \cap t_j\right|$ = the number of attribute values that are the same in t_i and t_j and $\left|t_i \cup t_j\right|$ = the number of attribute values in t_i or t_j. For example, $t_1 \cap t_4$ = {False} and $t_1 \cup t_4$ = {Sunny, Rainy, Warm, Mild, Soggy, Saturated, No, Yes, False}.

Then $\text{sim}(t_1, t_4) = 1/9$.

$t_1 \cap t_8$ = {Rainy, False, No} and

$t_1 \cup t_4$ = {Sunny, Warm, Mild, Soggy, Saturated, False, No}

Then $\text{sim}(t_1, t_8) = 3/7$.

There is a deficiency in the operation definition for the above formula. It assumes that attributes have an equivalent impact on similarity. However, if two tuples differ on an attribute having two values then the distance between them should be different to a higher degree than from the two other tuples which differ on an attribute having 100 values. The chances that two values are unequal are different for different attributes and are dependent upon the cardinality of the respective attribute value sets.

The following definition for similarity resolves the latter deficiency:

$$\text{sim}\left(t_i, t_j\right) = \frac{\left|t_i \cap t_j\right|}{\left|t_i \cap t_j\right| + 2\sum_{k \in t_i \cap t_j} \dfrac{1}{\left|D_k\right|}},$$

where D_k is the domain of an attribute k for which t_i and t_j have unequal values and k ranges over all such attributes for the two tuples. Consider t_1 and t_4 then D_1 = the domain set for Outlook and $\left|D_1\right| = 3$; D_2 = the domain set for Temperature and $\left|D_2\right| = 4$; D_3 = the domain set for Humidity and $\left|D_3\right| = 3$; and D_4 = the domain set for Play and $\left|D_4\right| = 2$.

Then $\sum_{k \in t_1 \cap t_4} \dfrac{1}{\left|D_k\right|} = \dfrac{1}{3} + \dfrac{1}{2} + \dfrac{1}{4} + \dfrac{1}{2} = 19/12$, $\left|t_1 \cap t_4\right| = 1$, and therefore $\text{sim}(t_1, t_4) = 6/25$.

Consider t_1 and t_8, then

D_1 = the domain set for Temperature and $\left|D_1\right| = 4$; D_2 = the domain set for Humidity and $\left|D_2\right| = 3$. Then $\sum_{k \in t_1 \cap t_8} \dfrac{1}{\left|D_k\right|} = \dfrac{1}{4} + \dfrac{1}{3} = 7/12$, $\left|t_1 \cap t_4\right| = 3$, and therefore $\text{sim}(t_1, t_4) = 18/25$.

Given a threshold, Θ, a value between 0 and 1, and a pair of objects T_i and T_j, then t_i and t_j are neighbors if $sim(t_i,t_j) \geq \Theta$. For example, let $\Theta = 0.20$. Then $sim(t_1,t_8) = \dfrac{6}{25}$ implies that t_1 and t_4 are neighbors. Also, sim $(t_1,t_8) = \dfrac{18}{25}$ implies that t_1 and t_8 are neighbors. Next, ROCK requires that the similarity of all pairs of tuples be found in order to determine the neighbors for each tuple and then for every pair of tuples find all their *common neighbors*. A sample computation is that $sim(t_1,t_2) = \dfrac{3}{3+2\left(\dfrac{1}{3}+\dfrac{1}{2}\right)} = 9/14$.

Using this computation process, the following vectors can be determined:

All $sim(t_1,t_i)$ for i = 1 to 15.   For all $sim(t_4,t_i)$ for i = 1 to 15.

	sim(t_1,t_i)
t_2	0.64
t_3	0.64
t_4	0.24
t_5	0.48
t_6	0.26
t_7	0.00
t_8	0.72
t_9	0.48
t_{10}	0.67
t_{11}	0.44
t_{12}	0.60
t_{13}	0.00
t_{14}	0.52
t_{15}	0.76

	sim(t_4,t_i)
t_1	0.24
t_2	0.24
t_3	0.62
t_5	0.57
t_6	0.264
t_7	0.26
t_8	0.64
t_9	0.69
t_{10}	0.72
t_{11}	0.80
t_{12}	0.80
t_{13}	0.80
t_{14}	0.48
t_{15}	0.58

Therefore t_1 has neighbors:

$$t_2, t_3, t_4, t_5, t_6, t_7, t_8, t_9, t_{10}, t_{11}, t_{12}, t_{13}, t_{14}, t_{15}.$$

When considering t_4 then every tuple is a neighbor of t_4.

As a result $\{t_2, t_3, t_4, t_5, t_6, t_8, t_9, t_{10}, t_{11}, t_{12}, t_{14}, t_{15}\}$ are common neighbors of t_4 and t_1.

A Link(t_i, t_j) between the tuples is defined as the number of *common neighbors* between t_i and t_j. A large value for the Link(t_i, t_j) indicates that t_i and t_j belong to the same cluster. Let D be the set of tuples. The link graph L for D is defined to be a graph with D as the set of vertices and with an edge between a pair of vertices t_i and t_j when Link(t_i, t_j) $\neq 0$.

Selection of the threshold, q, decides the density of the link graph. The higher the q value, the sparser the link graph. Then the graph L has a greater number of connected components if the q is chosen to be a high value.

Finally, the *goodness measure* $g(C_i, C_j)$ for merging two clusters C_i, C_j is defined as

$$g(C_i, C_j) = \frac{Link(C_i, C_j)}{\left[(n_i + n_j)^{1+2f(\Theta)} - n_i^{1+2f(\Theta)} - n_j^{1+2f(\Theta)}\right]},$$

where link(C_i, C_j) is the sum of the cross links between the tuples in C_i and C_j, and $f(\Theta) = (1 + \Theta)/(1 - \Theta)$.

The following explanation of the ROCK algorithm plus the preceding operational terms and the actual algorithm are from Dutta, Mahanta, and Pujari.[5]

The ROCK algorithm starts with each cluster being a single data point and keeps merging the pair of clusters with the best positive goodness measure. To determine the pair of clusters (C_i, C_j) having the highest goodness measure $g(C_i, C_j)$, a global heap Q and a local heap $q(C_i)$ for each cluster C_i are maintained. The local heap $q(C_i)$ contains each C_j with nonzero $g(C_i, C_j)$ and Q contains each cluster C_i with max_g($q(C_i)$), the maximum goodness measure in $q(C_i)$. The merging process is carried out until a specified number k of clusters remain or until the links between the clusters disappear, i.e., max_g(Q) becomes zero. The ROCK algorithm is described below:

Input: A set D of data-points

Number k of clusters to be found

The similarity threshold θ

```
begin
    compute nbrlist[i] for each i ∈ D using θ
    compute the links Link[i, j] for each i, j ∈ D
    for each x ∈ D
            build local heap q(x)
            build global heap Q
    while (size(Q) > k and max_g(Q) > 0)
            {u = extractmax(Q)
            v = max(q(u))
            delete(Q,v)
            w = merge(u,v)
                    for each x ∈ q(u) ∪ q(v)
                            {link[x,w] = link[x,u] + link[x,v]
                            delete(q(x), u);
                            delete(q(x), v);
                            insert(q(x), w, g(x,w));
                            insert(q(w), x, g(x,w));
                            update(Q, x, q(x))}
            insert(Q, w, q(w))
            }
    End
```

FIGURE 9.1 Algorithm ROCK.

9.3 STIRR

STIRR (**S**ieving **T**hrough **I**terated **R**elational **R**einforcement) is operationally defined as an iterative algorithm based on nonlinear dynamical systems. Let's start the study of STIRR by operationally defining what is meant by a dynamical system.

When using *dynamic programming* the problem to be solved is resolved by identifying a collection of subproblems and solving them one by one. Answers to these small problems lead from solutions to larger subproblems, until the all subproblems are solved. Dynamic programming problems have an underlying implicit directed graph, where the nodes are the subproblems needed to solve the original problem and graph's edges are the dependencies between the subproblems. In order to solve subproblem B, we need the answer to subproblem A, then there is an edge from A to B. In this case, A is thought of as a smaller subproblem than B and it will always be smaller.

Dasgupta, Papadimitriou, and Vazirani[6] provide the following example in Chapter 6 of their textbook:

> In the *longest increasing subsequence* problem, the input is a sequence of numbers $a_1, a_2, ..., a_n$. A *subsequence* is any subset of these numbers taken in order, of the form $a_{i1}, a_{i2}, ..., a_{ik}$ where $1 < i_1 < i_2 < ... < i_k < n$, and an *increasing* subsequence is one in which the numbers are getting strictly larger. The task is to find the increasing subsequence of greatest length. For instance, the longest increasing subsequence of 5, 2, 8, 6, 3, 6, 9, 7 is 2, 3, 6, 9.

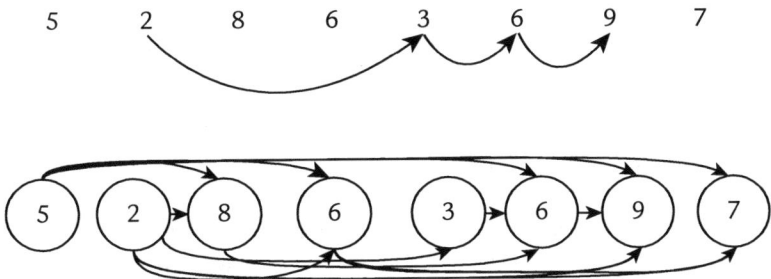

FIGURE 9.2 Increasing Subsequences.

In this example, the arrows denote transitions between consecutive elements of the optimal solution. More generally, to better understand the solution space, let's create a graph of *all* permissible transitions: establish a node i for each element ai, and add directed edges (i, j) whenever it is possible for a_i and a_j to be consecutive elements in an increasing subsequence, that is, whenever $i < j$ and $a_i < a_j$ (Figure 9.2).

Notice that: (1) this graph $G = (V, E)$ is a directed graph, because all edges (i, j) have $i < j$, and (2) there is a one-to-one correspondence between

⁶ Dasgupta, S., Papadimitriou, C. H., & Vazirani, U. V. (2007). *Algorithms*, McGraw-Hill.

increasing subsequences and paths in this dag. Therefore, our goal is simply to find the longest path in the dag!

Here is the algorithm:

```
for j = 1, 2, . . . , n:

    L(j) = 1 + max{L(i) : (i;j) ∈ E}

    return max_j L(j).
```

This is dynamic programming. In order to solve our original problem, we have defined a collection of subproblems $\{L(j) : 1 < j < n\}$ with the following key property that allows them to be solved in a single pass: (1) There is an ordering on the subproblems, and a relation that shows how to solve a subproblem given the answers to smaller subproblems, that is, subproblems that appear earlier in the ordering. In our case, each subproblem is solved using the relation $L(j) = 1 + \max\{L(i) : (i; j) \in E\}$, an expression that involves only smaller subproblems. How long does this step take? It requires the predecessors of j to be known; for this the adjacency list of the reverse graph GR, constructible in linear time is handy. The computation of $L(j)$ then takes time proportional to the in-degree of j, giving an overall running time linear in $\{E\}$. This is at most $O(n^2)$, the maximum being when the input array is sorted in increasing order. Thus the dynamic programming solution is both simple and efficient.

Dynamic programming takes advantage of the duplication and arrangement to solve each subproblem only once. Additionally, it saves the solution (in a table or in a globally accessible place) for later use. The underlying idea of dynamic programming is: avoid calculating the same stuff twice, usually by keeping a table of known results of subproblems. Dynamic programming is a tableau method. Unlike divide-and-conquer, which solves the subproblems top-down, dynamic programming uses a bottom-up technique. The dynamic programming technique is related to divide-and-conquer, in the sense that it breaks the problem down into smaller problems and it solves recursively.

STIRR operationally defines a database as a set of tuples. The algorithm employed is a weight-propagation method. First, an item of interest is seeded, or assigned, a small weight. Then items associated with the item of interest are assigned weights, thus the weight of the item of interest has propagated to associated items. Then the associated items propagate weights further. Note that items highly related to the item of interest acquire weight and the propagation of the weights is transitive. This weight propagation process is a nonlinear dynamic system derived from a table of categorical data.

The database is a graph where each distinct value in the domain of each attribute is represented by a weighted node. For each tuple in the database, an edge represents a set of nodes which participate in that tuple. Each tuple in the database is represented with attribute values as a node and edges are to represent connections between the attribute values for the specific tuple.

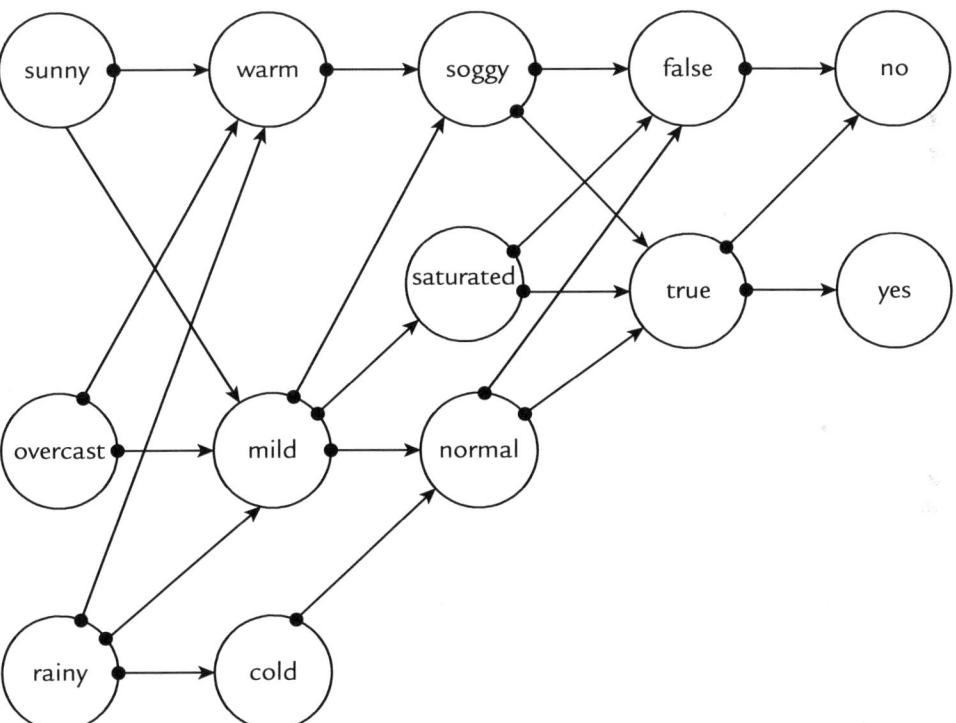

FIGURE 9.3 Representation of the Weather Data Tuples.

Gibson, Klienberg, and Raghavan[2] define a *configuration* as an assignment of a weight w_v for each node v. Next the weights are normalized, or the sum of the squared weights is equal to one, accomplished through rescaling. Repeated application of a function on the weights is performed until a fixed point is reached, a point u is found where $f(u) = u$. Then no more weight-propagation could take place from the fixed point further. The function chosen is a *combiner* function, Ø. Then, the following algorithm is performed:

For each tuple $t = \{v, u_1, u_2,..., u_{n-1}\}$,

containing v do,

$$x_t \leftarrow \phi(u_1, u_2,..., u_{n-1}),$$

$$w_v \leftarrow \sum_t x_t.$$

FIGURE 9.4 Weight Update Algorithm, w_v.

This algorithm simply updates the weight of v by updating it on all members of the set of tuples that contain v. To complete each step in the iteration, the function f is executed by updating the weight of each w_v and then normalizing the set of weights. The end of each iteration step generates a new configuration $f(w)$.

To ensure that the combiner function, Ø, is simple and easy to use implies that one chooses either the product function or the sum function for the combiner function. Note that the sum function is linear with respect to the tuples. The product function and $S_p(w_1, w_2,..., w_k) = (w_2^p, w_2^p,..., w_k^p,)^{(1/p)}$ involves a nonlinear term for each tuple, that enables a greater potential to encode co-occurrences within tuples.

Gibson, Kleinberg, and Raghavan[2] have experimentally determined that the weight sets generally converge to fixed points or the weight sets converge to cycles through a finite set of values. These final configurations, due to cycles through a finite set of values, are referred to as a *basin*. The convergence is dependent on the *combiner function*. Analyzing the stability is hard for any arbitrary combiner function. However, for simple combiner functions like sum or multiplication, the system definitely converges to a fixed point. It is easy to see that for categorical attributes, the values that are related through common tuples influence each other during weight modification. Thus one does not really require any similarity metric to be defined for categorical attributes. Interestingly, in order to cluster the set of tuples, STIRR maintains multiple copies of weights. When the fixed point is reached, the weights in one or more of the basins isolate two groups of attribute values on each attribute—the first with large positive weights and the second with small negative weights. The nodes with large positive weights and large negative weights are grouped to determine clusters. These groups correspond intuitively to projections of clusters on the attribute. However, the automatic identification of such sets of closely related attribute values from their weights requires a nontrivial postprocessing step; such a postprocessing step was not addressed in the work by Gibson, Kleinberg, and Raghavan.[2] Moreover, the postprocessing step will also determine what "clusters" are output. The underlying idea of STIRR is unique but it may be hard to analyze the stability of the system for any useful combiner functions. One requires rigorous experimentation and fine tuning of parameters to arrive at a meaningful clustering. One method of setting the initial configuration is to set small weights to 1 and then normalize the weights, this is the uniform initialization. For a randomized initialization, each weight is set to an independently random value in [0, 1] and then the random weights are normalized.

9.4 CACTUS

CACTUS by Ganti, Gehrke, and Ramakrishnan[2] emphasizes the concept of *strongly connected sets*. In this method, a set of objects, $C = \{C_1, C_2, \ldots, C_n\}$ is called a cluster if:

1. For all $i, j \in \{1, 2, \ldots, n\}$ where $i \neq j$, C_i and C_j are strongly connected.

2. For all $i \in \{1, 2, \ldots, n\}$ C_i is maximal.

3. The support fulfills $\sigma(C) > \alpha|D|$, where $\sigma(C)$ is the support for cluster C and α is a threshold constant.

Two sets of values $C_1 \subseteq D_i$ and $C_2 \subseteq D_j$ of different attributes are called *strongly connected*, if all pairs of values $a_i \in D_i$ and $a_j \in D_j$ occur more frequently than expected.

The CACTUS is composed of three components: the summarization phase, the clustering phase, and the validation phase. Inter-attribute and intra-attribute summaries are formulated by the summarization as well as accessing the data. During the clustering phase, cluster candidates are determined. The validation is when the actual clusters are chosen from the set of candidates.

CACTUS (**C**lustering **C**ategorical Data **U**sing **S**ummaries)[7] is a sort of subspace clustering. CACTUS attempts to split the database vertically and tries to cluster the set of projections of these tuples to only a pair of attributes. Its basic principle can be described as follows. Consider two attribute values of two different attributes in the database. Say, a_i of attribute type A and a_j of attribute type B. There may be tuples where a_i and a_j co-occur (i.e., occur together). The support of these two values in the database is the proportion of tuples in which they appear together. If this support exceeds a prespecified value, we say that these values are strongly connected. This concept can be used to compute the inter-attribute and intra-attribute summaries of the given data set. Most interesting aspects of these steps are that these can be computed using inter-attribute and intra-attribute summary. It is not necessary to refer to the original data base. CACTUS first identifies the cluster projections on all pairs of attributes by fixing one attribute. Then it generates an intersecting set to represent the cluster projection on this attribute for n-cluster (involving all the attributes). Once all the cluster projections on individual attributes are generated, these are synthesized to get the clusters of the database. Thus the major steps of CACTUS are:

1. Finding cluster projections on a given attribute A_i with respect to another attribute A_j.

2. Intersecting all the cluster projections for any given A_i to get the cluster projection of A_i with respect to all the attributes.

3. Synthesizing the resulting cluster projections to get the main clusters.

In order to illustrate the CACTUS algorithm, several terms need to be operationally defined. These include the *support for an attribute value pair*, how attribute value pairs and two sets of attribute values are *strongly connected*, a *cluster over a set of attributes*, the *similarity between two attribute values*, an *inter-attribute summary*, and an *intra-attribute summary*. The discussion will be centered on Figure 9.2.

Let A_1, A_2,..., A_n be a set of attributes with domains D_1, D_2,..., D_n. Let $t \in \{D_1 x D_2 x \ldots x D_3\}$ represent the tuples in the data set. The *support* $\sigma_D(a_i, a_j)$ for two attribute value pairs is the number of tuples in the data set with $t.A_i = a_i$ and $t.A_j = a_j$. In Figure 9.3, $\sigma_D(sunny, warm) = 1$ because the fact that there is only one entry point into the "warm" node and one exit point only allows for one tuple to contain "sunny" and "warm." In a similar fashion, $\sigma_D(sunny, mild) = 9$ because "sunny" has one entry to "mild," and "mild" has 6 paths leading to "no" and 3 paths leading to "yes." Note that $\sigma_D(mild,true) = 9$ because there are 3 entry points to "mild" and 3 paths leading to "true" from "mild."

The expected support for an attribute pair a_i and a_j is $E\left[\sigma_D(a_i, a_j)\right] = \dfrac{|D|}{|D_i| x |D_j|}$, based upon the assumption that all the attributes are independent and attribute pairs are equally likely. For the weather data set $|D| = 15$. Additionally, $|D_1| = 3$, $|D_2| = 4$, $|D_3| = 3$, $|D_4| = 2$, and $|D_5| = 2$. Then $E\left[\sigma_D(sunny, mild)\right] = \dfrac{15}{3x4} = 1.25$ and $E\left[\sigma_D(mild, true)\right] = \dfrac{15}{4x2} = 1.88$.

Let a_i and a_j be 2 attribute values and $a > 1$. This pair of attribute values is *strongly connected* with respect to D if $\sigma_D(a_i, a_j) > a\dfrac{|D|}{|D_i| x |D_j|}$ and

$$\sigma_D^*\left(a_i, a_j\right) = \begin{cases} \sigma_D\left(a_i, a_j\right), & \text{if } a_i \text{ and } a_j \text{ are strongly connected} \\ 0, \text{otherwise}, & \text{where } i \neq j. \end{cases},$$

An attribute value a_i is *strongly connected* to a subsect S_j of D_j if a_i is *strongly connected to every* a_j in S_j. Two subsets, S_i and S_j of D, a set of tuples itself and represented as a set of attribute values, are strongly connected if every value in S_i is strongly connected to every value in S_j and every value in S_j is strongly connected to every value in S_i. For the example in Figure 9.3 let $a = 2$. Then $\sigma_D(sunny, warm) = 1$ and $2\dfrac{|D|}{|D_1| x |D_2|} = 2.5$. Therefore "sunny" and "mild" are not strongly connected. However, $\sigma_D(sunny, mild) = 3$ and then "sunny" and "mild" are strongly connected. Notice that $\sigma_D(mild, ture) = 9$ and $2\dfrac{|D|}{|D_2| x |D_4|} = 2\left\{\dfrac{15}{4x2}\right\} = 3.75$ implies that "mild" and "true" are strongly connected.

For $i = 1, 2, \ldots, n$, let C_i be a subset of D_i with 2 or more members and $a > 1$. Then $C = \{C_1, C_2, \ldots, C_n\}$ is a cluster over $A_1, A_2, \ldots, A_n$ if: C_i and C_j are strongly connected for $i,j \in [0,1]$ and $i \neq j$. For all $i,j \in [0,1]$ and $i \neq j$ there does not exist a super set C_i' of C_i where for every $j \in [0,1]$ and $i \neq j$ that C_i and C_j are strongly connected. The support $\sigma_D(C)$ of C is at least at times the expected support of C under the assumption that the attributes are independent and the attribute values in each attribute are equally likely. The support of C is the number of tuples in D that belong to C, or the number of all tuples in D where $t.A_i \in C_i$. Computations for the support for various attribute value pairs include:

$\sigma_D(sunny,warm) = 3$; $\sigma_D(overcast,warm) = 3$; $\sigma_D(rainy,warm) = 3$;

$\sigma_D(sunny,mild) = 8$; $\sigma_D(overcast,mild) = 8$; $\sigma_D(rainy,mild) = 8$;

$\sigma_D(sunny,cold) = 0$; $\sigma_D(overcast,cold) = 0$; $\sigma_D(rainy,cold) = 3$;

infer that D_1 and D_2 are not strongly connected because $\sigma_D(sunny, cold) = 0$. D_1 and $\{mild\}$ are strongly connected. The additional computations:

$\sigma_D(mild,soggy) = 5$; $\sigma_D(soggy,false) = 15$; $\sigma_D(saturated,false) = 3$;

$\sigma_D(mild,saturated) = 5$; $\sigma_D(soggy,true) = 10$; $\sigma_D(saturated,true) = 6$;

$\sigma_D(mild,normal) = 5$; $\sigma_D(normal,false) = 3$; $\sigma_D(normal,true) = 6$;

$\sigma_D(false,no) = 11$; $\sigma_D(false,yes) = 0$; $\sigma_D(mild,yes) = 9$;

$\sigma_D(mild,no) = 12$; $\sigma_D(overcast,soggy) = 6$; $\sigma_D(overcast,saturated) = 1$;

$\sigma_D(overcast,normal) = 3$; $\sigma_D(rainy,soggy) = 6$; $\sigma_D(rainy,saturated) = 3$;

$\sigma_D(rainy,normal) = 6$;

lead to the cluster: $C_1 = \{sunny,rainy\}$, $C_2 = \{mild\}$, $C_3 = \{soggy,normal\}$, $C_4 = \{true\}$, $C_5 = D_5$. To capture the clusters, both a similarity measure and summaries need to be operationally defined.

Let a_1 and a_2 be members of the same domain of attribute values D_i; Then the similarity function γ^j is defined as: $\gamma^j(a_1, a_2) = \left\| \left\{ x \in D_j \mid i \neq j, \sigma_D^*(a_1, x) > 0 \text{ and } \sigma_D^*(a_2, x) > 0 \right\} \right\|$. Note that $\gamma^2 (sunny,rainy) = 2$.

Let $A_1, A_2,..., A_n$ be a set of categorical attributes with domains D_1, $D_2,..., D_n$. Let D be the associated data set of tuples. The *inter-attribute summary*, Σ_{ij} is defined as:

$$\sum_{ij} = \left\{ (a_i, a_j, \sigma_D^*(a_i, a_j)) \mid i, j \in [0, 1], i \neq j, a_i \in D_i, a_j \in D_j, \text{ and } \sigma_D^*(a_i, a_j) > 0 \right\}.$$

Note that $\Sigma_{12} = \{(sunny,warm, \sigma_D^*(sunny,warm) = 3), (sunny,mild, \sigma_D^*(sunny, mild) = 8), (overcast,warm, \sigma_D^*(overcast,warm) = 3), (overcast, mild, \sigma_D^*(overcast,mild) = 8), (rainy,warm, \sigma_D^*(rainy,cold) = 3)\}$.

The *intra-attribute summary*, Σ_{ii} is defined as:

$$\Sigma_{ii}^{j} = \{(a_{i1}, a_{i2}, \gamma^{j}(a_{i1}, a_{i2})) | i1, i2 \in [0, 1], i1 \neq i2, a_{i1} \in D_{i}, i2 \in D_{i}, \text{ and } \gamma^{j}(a_{i1}, a_{i2}) > 0\}.$$

Then

$(\Sigma_{11})^{2}$ = {(*sunny,overcast*, γ^{2} (*sunny,overcast*) = 2), (*sunny,rainy*, γ^{2}(*sunny,rainy*) = 2), (*overcast,rainy*, γ^{2}(*overcast,rainy*) = 2)}.

In the summarization phase of the CACTUS algorithm, for every attribute value pair a counter is set to zero. Next, the data set is scanned. For each tuple, the counter is incremented for the pair $(t.A_{i}, t.A_{j})$, $t \in D$. At the end of the scan, compute $\sigma_{D}^{*}(a_{i}, a_{j})$ for each attribute value pair a_{i} and a_{j} by setting all counters to zero whose value is less than the threshold $k_{ij} = 2\dfrac{|D|}{|D_{i}| x |D_{j}|}$. This process counts only the strongly connected attribute value pairs. Then the strongly connected pairs can be retained in a matrix format for storing sparse matrices. This matrix can be modified to have each cell hold the triple for the inter-attribute summaries, which can be retrieved by scanning the matrix.

Using the following Structured Query Language (SQL) the following statement joins Σ_{ij} with itself to compute the set of attribute value pairs of A_{2} strongly connected to each other with respect to A_{j}.

Select	$t_{1}.A, t_{2}.A, \text{count}(^{*})$
From	$\Sigma_{ij} \text{ as } t_{1}(A,B), \Sigma_{ij} \text{ as } t_{2}(A,B)$
Where	$t_{1}.A \neq t_{2}.A \text{ and } t_{1}.B \neq t_{2}.B$
Group by	$t_{1}.A, t_{2}.A$
Having	$\text{count} > 0;$

The intra-attribute summaries can then be computed at any time by application of the SQL statement.

The clustering phase consists of two steps. First, an analysis of each attribute is run to compute all cluster-projections on it. Each C_{i} in a cluster C is a cluster-projection on A_{i}. Second, a determination of candidate clusters on a pair of attributes, and then extends the pair to a set of three attributes, and so on. This is a level-wise synthesis of candidate clusters on sets of attributes from the cluster-projections on individual attributes.

The validation phase is also based upon a scan. The candidate clusters are scanned and any cluster that fails the threshold condition have their support deleted.

9.5 CLICK

CLICK (CLIque Clustering using K-partite graphs) uses a graphical approach for partitioning the categorical data. Therefore, a quick review of graphics is helpful in fully understanding of the CLICK algorithm.

A graph is a triple $(V(G), E(G), R)$ with a set of vertices, $V(G)$, a set of edges, $E(G)$, and a relation $R:E(G) \rightarrow V \times V$.

A graph, G, is *connected* if there is a path in G between any pair of vertices.

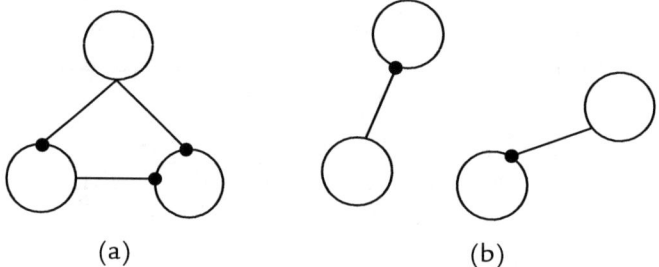

(a)   (b)

FIGURE 9.5 (a) A Connected Graph, and (b) A Disconnected Graph.

For a graph G with vertex set $V(G)$ and edge set $E(G)$, a *subgraph* is a graph whose vertex set is a subset of $V(G)$ and edge set is a subset of $E(G)$.

A *complete graph* is a graph in which every distinct pair of edges is represented by only one edge. Note that a complete graph with n vertices must possess exactly $n(n-1)/2$ edges. Figure 9.6 illustrates some complete graphs.

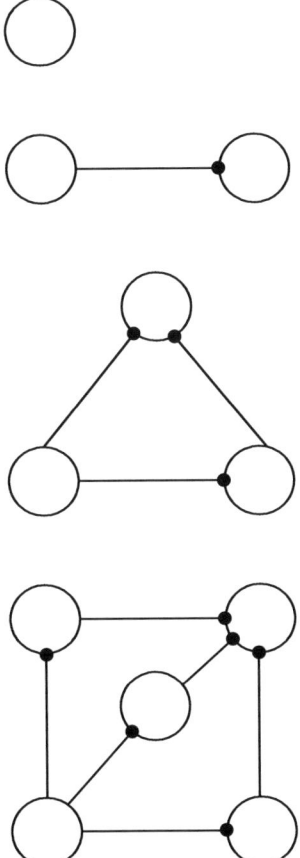

FIGURE 9.6 Complete Graphs.

A K-partite graph is a graph where the vertex set can be split into K sets, $\{A_1, A_2, ..., A_k\}$ in a manner where each edge of the graph joins a vertex in $v_1 \in A_i$ and $v_2 \in A_j$, $i, j \in [1, K]$, and $i \neq j$.

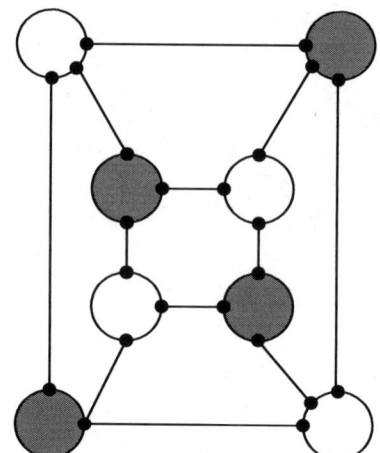

FIGURE 9.7 A Bipartite Graph.

Consider the following graph *G*.

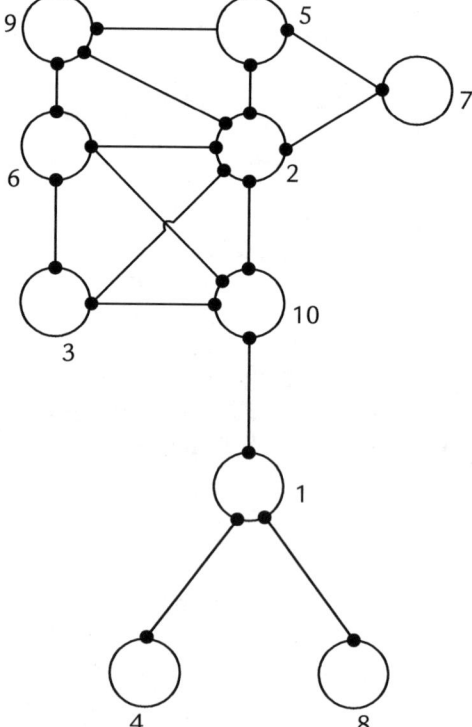

FIGURE 9.8 Sample Study Graph.

This graph can be represented by the following adjacency matrix $A = (a_{i,j})$ where $a_{i,j}$ is 1 if vertices i and j are joined by an edge or is 0 otherwise.

0	0	0	1	0	0	0	1	0	1
0	0	1	0	1	1	1	0	1	1
0	1	0	0	0	1	0	0	0	1
1	0	0	0	0	0	0	0	0	0
0	1	0	0	0	0	1	0	1	0
0	1	1	0	0	0	0	0	1	1
0	1	0	0	1	0	0	0	0	0
1	0	0	0	0	0	0	0	0	0
0	1	0	0	1	1	0	0	0	0
1	1	1	0	0	1	0	0	0	0

TABLE 9.3 Adjacency Matrix.

CLICK uses an algorithm, similar to the one in Figure 9.9, to determine the maximal complete subgraphs of a graph. The algorithm constructs a new matrix, $B = (b_{i,j})$ where $b_{i,j}$ equals 1 if vertex i is in the maximal complete subgraph j and is 0 otherwise. Each column of B identifies a maximal complete subgraph.

Let *NR* be the number of rows

Let *NK* be the maximal complete subgraph number

Let *UBK* be the upper bound for *NK*

Let *NC* be the number of columns

Input: $A_{NR,NC}$ is the adjacency matrix for the graph

Output: $C_{NR,NC}$ is the maximal complete subgraph identification matrix

$C_{1,1} \leftarrow 1$

$UBK \leftarrow 1$

```
Begin {Algorithm}

For NR = 2 to n
```

```
Begin
 For NK = 1 to UBK
        Begin
                C_{NR,NK} ← 1
        End NK
 For NC = 1 to (NR − 1)
        Begin
                If a_{NR,NC} = 0 then
                        For NK = 1 to UBK
                                If C_{NC,NR} =1 then C_{NR,NK} ← 0
                        End NC
 For NC = 1 to (NR − 1)
        Begin
                If a_{NR,NC} = 0 then
                        For NK = 1 to UBK
                                Begin
                                        If C_{NC,NK} C_{NR,NK} = 2
                                                then C_{NC,NK} ← 0
                                End NK
        End NC
 For NC = 1 to (NR − 1)
        Begin
                If a_{NR,NC} = 1 then
                        If there exists NK = 1 to UBK where
                                C_{NC,NK}C_{NR,NK} = 1 then
                                        UBK ← UBK + 1
        End NC
 End NK
```

If there exists $NK = 1$ to UBK where $C_{NC,NK}C_{NR,NK} = 2$ then

  Begin

$$CS_{NC,NK} \leftarrow 1$$

For $m = 1$ to $(NR - 1)$

    If $C_{NC,NK}C_{NR,NK} = 2$ and $a_{NC,m} = 0$ then $C_{m,NK} \leftarrow 0$

For $m = 1$ to $(NR - 1)$

$$C_{m,UBK} \leftarrow 2$$

$$C_{NR,UBK}C_{NC,UBK} \leftarrow 1$$

For $m = 1$ to NC

    If $C_{m,NK} = 2$ then

      for $m_1 = 1$ to NR

        If $C_{m1,UBK} = 1$ and $a_{m,m1} = 0$ then

$$C_{m,UBK} = 0$$

  End m

For $NK = 1$ to UBK

  Begin

    For $NC = 1$ to NR

      Begin

        If $C_{NC,NK} = 2$ then

        Begin

$$C_{NC,NK} \leftarrow 1$$

          For $m = 1$ to NR

            Begin

            If $C_{NC,NK}C_{m,NK} = 2$ and

$$a_{NC,m} = 0 \text{ then}$$

$$C_{m,NK} \leftarrow 0$$

$$NC = NC - 1$$

            End m

      End NC

    End NK

  End {algorithm}

FIGURE 9.9 Clique Algorithm.

1	0	0	1	0	0	1
0	1	1	0	1	1	0
0	1	0	0	0	0	0
1	0	0	0	0	0	0
0	0	1	0	1	0	0
0	1	0	0	0	1	0
0	0	1	0	0	0	0
0	0	0	1	0	0	0
0	0	0	0	1	1	0
0	1	0	0	0	0	1

TABLE 9.4 The Maximal Complete Subgraphs for Graph G.

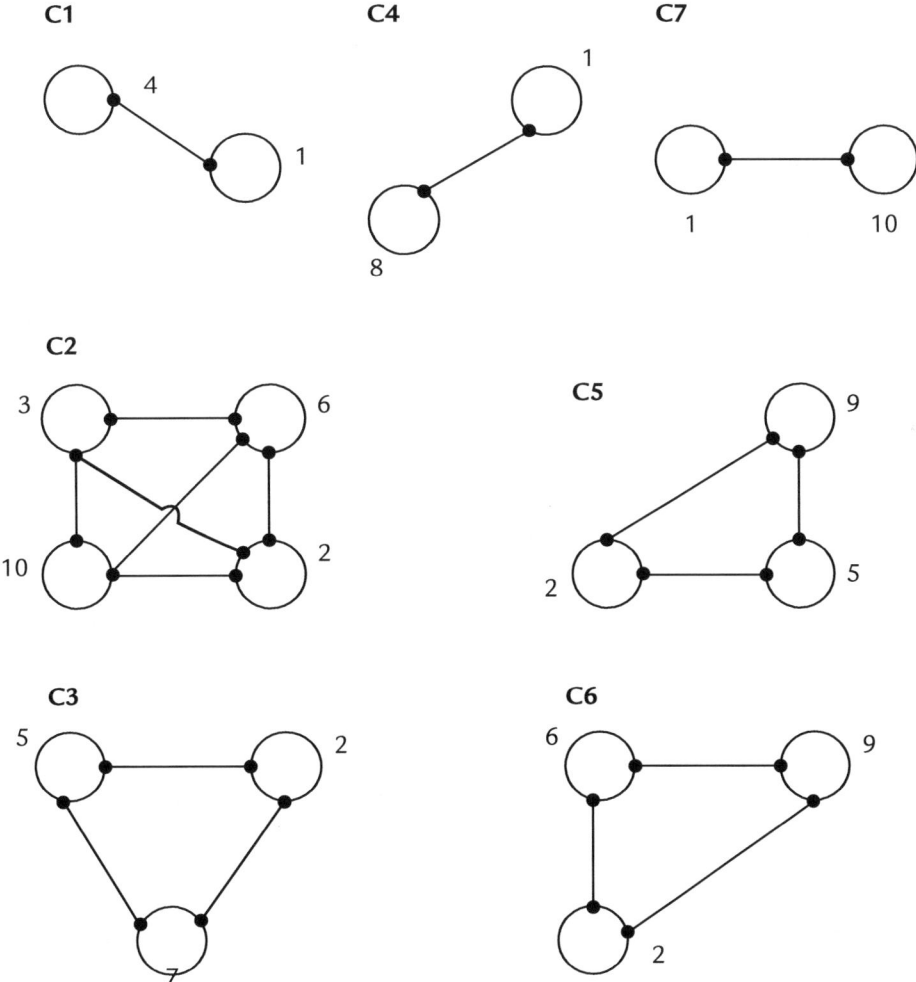

FIGURE 9.10 The Maximal Complete Subgraphs for Graph G.

The CLICK algorithm is composed of three modules: pre-processing, clique (maximal complete subgraphs) detection, and post-processing. Pre-processing constructs the clique matrix from the adjacency matrix, which is derived from the input data set. Note that the attributes are ranked for efficiency reasons, using the strongly connected attribute values concept. The clique detection module identifies all the k-partite cliques using an algorithm similar to Figure 9.10. During this module, care is taken to maintain strongly connected cliques. Post-processing utilizes the support of the candidate cliques, identified in the clique detection module, to find the final clusters. These final clusters are optimally merged to partially relax the strict cluster conditions. Post-processing is accomplished in a single scan.

9.6 SUMMARY

- The ROCK method defines two data points to be similar if they tend to have a large number of common neighbors.

- A data point belonging to a cluster C_i has $n_i^{f(\Theta)}$ neighbors in C_i where

$$f(\Theta) = \frac{1 - \Theta}{1 + \Theta}.$$

- ROCK is a hierarchical clustering algorithm based upon merges of clusters possessing maximal goodness measure.

- STIRR method is a hypergraph method based upon utilizing dynamic programming, where weights are assigned to each node.

- Two sets of attribute values for distinct attribute domains are strongly connected, if all pairs of values in the respective attribute domains occur more frequently than expected.

- CACTUS builds inter-attribute and intra-attribute summaries for determining cluster candidates chosen on the basis of being maximal strongly connected clusters with support greater than a predefined threshold, a function of the total number of tuples in the data set.

- CLICK constructs a matrix representation of the maximal complete subgraphs of the data set's adjacency matrix.

9.7 EXERCISES

Use the following data set, a subset of the Car Evaluation Database accessible from the UCI Learning Laboratory, to complete the Exercises 1 through 10.

Buying	Maint	Doors	Persons	Lug_Boot	Safety	Class
Low	high	2	more	big	Med	Acc
Low	high	2	more	big	High	Vgood
Vhigh	med	4	more	med	Med	Acc
Vhigh	vhigh	2	2	small	Low	Unacc
Vhigh	vhigh	2	2	small	Med	Unacc
Vhigh	vhigh	2	more	small	High	Unacc
Vhigh	high	2	4	small	Med	Unacc
High	med	2	more	Small	High	Unacc
High	med	2	more	Med	Low	Unacc
Low	vhigh	2	2	Small	Med	Unacc
Low	vhigh	2	2	Small	High	Unacc
Low	low	5 more	more	Med	Low	Unacc
Low	low	5 more	more	Med	Med	Good

TABLE 9.5 Auto Purchasing Data.

The attribute values are:

Buying	vhigh, high, med, low
Maint	vhigh, high, med, low
Doors	2, 3, 4, 5 more
Persons	2, 4, more
lug-boot	small, med, big
Safety	low, med, high
purchase	unacc, acc, good, vgood

1. Represent this database as a graph, based upon attributes are weighed nodes. Each tuple in the database is represented with attribute values as a node and edges are to represent connections between the attribute values for the specific tuple.

2. Find the similarity between the second and fifth tuples using both formulas discussed in the section on ROCK.

3. For the above data set, find the common neighbors for the second and fifth tuples where $\Theta = 0.2$ and $\Theta = 0.61$.

4. Using the above data set, find and interpret the results for the Link (second tuple, fifth tuple) value for $\Theta = 0.2$ and for $\Theta = 0.61$.

5. Determine the goodness measure for two singleton clusters $\{t_2\}$ and $\{t_5\}$ for the above data set for $\Theta = 0.2$ and for $\Theta = 0.61$.

6. Perform a structured walk-thru for the ROCK algorithm applied to the following data set:

Buying	Maintenance	Safety	Class
Low	High	Medium	Acceptable
Low	High	High	Very good
Very high	Medium	Medium	Acceptable
Very high	Very high	Low	Unacceptable
Very high	Very high	Medium	Unacceptable
Low	Low	Medium	Good

TABLE 9.6 Auto Purchasing Data Subset One.

7. Implement the ROCK algorithm as a computer program and run the program on the actual Car Evaluation Database in the UCI Learning Laboratory or the subset of the database.

8. Using dynamic programming, solve the following problem: Devise an algorithm, using a table, for paying a given amount to a customer using the smallest possible number of coins.

9. Let G be a directed graph with a set N of nodes and a set E of edges. Each edge has an assigned non-negative length. Use dynamic programming to find the length of the shortest path between each pair of nodes.

10. Provide an outline, in your own words, of how the STIRR method would solve the clustering for the subset of the Car Evaluation Database.

11. Determine the support values for the attributes given the following database:

Maintenance	Safety	Class
High	Medium	Acceptable
High	High	Very Good
Low	Medium	Unacceptable
Low	Medium	Very Good
High	Medium	Unacceptable
Low	High	Acceptable

TABLE 9.7 Auto Purchasing Data Subset Two.

12. For $a = 1.5$ find the pairs of attributes that are strongly connected for the database in Exercise 11. Redo this exercise for $a = 3$.

13. Find the inter-attribute summaries for the database in Exercise 11.

14. Find the intra-attribute summaries for the database in Exercise 11.

15. Encode the graph for Exercise 1 as an adjacency matrix for the database in Exercise 11.

16. Find the maximal complete subgraphs for the graph in Exercise 11 using the algorithm in Figure 9.10.

MINING OUTLIERS

In This Chapter

10.1 INTRODUCTION

Data and information consumers expect their data sets, or large databases, to be accurate. However, database management systems rarely, or sometimes not at all, provide a specified level of accuracy for the database. The occurrence of inaccurate information eventually raises the need for data quality control. High cost, time, and personnel requirements are all involved in attempts to repair an existing database. Data cleansing is the needed activity for achieving data quality.

Many varieties of situations exist for dirty data including:

1. Sanity checking was not performed for numeric input.

2. Time driven data is now out-of-date.

3. Business rules were not programmatically enforced, with triggers or stored procedures, or have changed since the system was put into production.

4. Field domain values have changed.

5. Data migration into a new system, or merging of systems allowed noncompliant values.

6. The presence of outliers is causing informational inaccuracy.

When the latter problems are present in the data, then the objective is to identify transactions that contain the errors. Through the use of specialized personnel, actions to manually correct the errors will be available. An effort needs to be cost-effective in outlier detection and data cleaning.

Hawkins[1] defines: "An outlier is an observation which deviates so much from other observations as to arouse suspicions that it was generated by a different mechanism." The objective of this chapter is to study methods meeting the goal of outlier detection, or to identify and understand the different mechanism. Statistical methods, clustering, and fuzzy clustering are the outlier detection methods discussed in this chapter.

10.2 OUTLIER DETECTION METHODS

Knorr, Ng, and Tucak[2] discuss algorithms and applications of *distance-based outliers*. A distance-based outlier is a datum for which at least p percent of the data possesses a distance larger than a specified distance. When the Euclidean distance function is the measure of dissimilarity, then the outliers are the points that can be isolated from the rest of the data set by a neighborhood, or ball, with a radius of the specified distance. The distance-based outlier algorithms are difficult to work with due to lack of being given the specification of the required distance and probability of the data set with the same required distance, for isolating each of the points with a neighborhood. Also distance-based outliers do not provide a ranking scheme. Distance-based outlier algorithms include nested loop algorithms, index-based algorithms, and partition-based algorithms.

Besides being separable from the data set many times in an application, there are instances where the outliers are part of the overall data set but

---

[1] Hawkins, D. (1980). *Identification of Outliers*. London: Chapman and Hall.

[2] Knor, E. M., Ng, R. T., and Tucakov, V. (2000). Distance-based outliers: Algorithms and Applications. *VLDB Journal,* 8(3-4), 237-253.

are distinct from their local neighborhood of data as addressed by Breunig, Kriegel, Ng, and Sander.[3] A measurement of the density of a neighborhood that captures the local data points and simultaneously excludes the local outlier needs to be determined. The determination of this local reachability distance can require large run times. With the aid of clustering methods, the local outliers can be detected by two disjoint neighborhoods; one containing the local neighborhood of data points and the other containing only the outlier. The distance between the two neighborhoods then represents the degree of separation of the local outlier, therefore a ranking of outliers is possible.

Brieunig, Kriegel, Ng, and Sander[4] discuss a *Local Outlier Factor method* created by Tan, Steinbach, and Kumar, based on scoring outliers on the basis of the density in the neighborhood. The Outlier Factor method uses the following definition:

The outlier score of an object is the reciprocal of the density in the object's neighborhood, where density is the average distance to the k nearest neighbors:

$$\text{density}(x,k) = \left(\frac{\sum_{y \in N(x,k)} distance(x,y)}{|N(x,k)|} \right)^{-1}.$$

A different density approach is to locate outliers using the relative density of points.

The relative density of a point x is the ratio of the density of a point and the average density of its neighbors:

$$\text{reldensity}(x,k) = \frac{density(x,k)}{\sum_{y \in N(x,k)} density(y,k)/|N(x,k)|}$$

The outlier score of a point is defined to be its relative density. Besides selecting the right k is difficult, the algorithm requires quadratic computation time.

---

[3] Brieunig, S., Kriegel, H. P., Ng, R., & Sander, J. (September, 1999). Optics of identifying local outliers, in *Lecture Notes in Computer Science*, (1704), 262-280.

[4] Brieunig, M. M., Kriegel, H. P., Ng, R. T., & Sander, J. (2000). Lof: identifying density-based local outliersl. In *Proceedings of the 2000 ACM SIGMOD International Conference on Management of Data*, Dallas, Texas, USA, 93-104.

This brief discussion has only touched the surface of outlier types. The interested reader should consult: Yu and Aggarwal[5] for discussion on *subspace outliers*, Hinneburg, Keim, and Wawryniuk[6] for discussion of *projection outliers*, review literature on *sequential outliers*, and MINDS. MINDS, the Minnesota Intrusion Detection System discussed by Lazarevic, Ertoz, Kumar, Ozgur, and Srivastava[7] uses the LOF.

10.3 STATISTICAL APPROACHES

A good starting point for finding distance-based outliers is the application of the empirical rule or Chebychev's inequality. The question being answered is: "What is the probability that a point will lie within k standard deviations of the mean?" If the Central Limit Theorem is invoked, then the sampling distribution of the sample mean is normally distributed with mean μ and standard deviation σ. The question is reduced to: "What is the area under the normal probability density function from $(\mu - k\sigma)$ to $(\mu + k\sigma)$?" The answer is: 0.997 if $k = 3$, 0.954 if $k = 2$, and 0.683 if $k = 1$. The answer to the original question is that the probability of finding a point that is more than three standard deviations away from the mean, an outlier, is less than 0.003. If the population distribution is unknown, then Chebychev's inequality can be used to answer the question. Chebychev's inequality states that for any population distribution at least $[1 - (1/k)^2](100)\%$ of the data points lie within k standard deviations of the mean. Therefore, the probability of finding a point, in an arbitrary distribution, which is more than 3 standard deviations from the mean, an outlier, is less than $1/9 = 0.11$.

Another useful result from statistics is the following theorem by Chernoff: Let $X_1, X_2, \ldots, X_n$ be a sequence of identically independently distributed 0-1 random variables where $p = P(X_i = 1)$, $X = \sum_i X_i$ and $\mu \geq E[X]$. Then for any $\delta > 0$

$$P(X > (1 + \delta)\mu) < \left[\frac{e^\delta}{(1 + \delta)^{1 + \delta}} \right]^\mu$$

---

[5] Aggarwal, C. C., & Yu, P. S. (2001). Outlier detection for high dimensional data, In *Proceedings of the 2000 ACM SIGMOD International Conference on Management of Data*, Santa Barbara, California, USA.

[6] Hinneburg, A., Keim, A., & Wawryniuk, M. (2003). Using projections to visually cluster high-dimensional data, *IEEE Computing in the Science of Engineering*, 5(2): 14-25.

[7] Lazarevic, A., Ertoz, L., Kumar, A., Ozgur, A., & Srivastava, J. (2003). A comparative study of anomaly detection schemes in network intrusion detection, In *SDM*.

enables the computation of finding the probability of a data point being in the region of rejection of a confidence interval, or being identified as a local outlier. In this case:

$$P(X \leq (1-\epsilon)np) \leq e^{\frac{-\epsilon^2 np}{2}} \text{ and } P(X \geq (1-\epsilon)np) \leq e^{\frac{-\epsilon^2 np}{2}}.$$

Note that these bounds are not symmetric.

Shawne-Taylor and Cristianini[8] discuss the *Hoeffding Like Weak Inequality* which states: Let $\{x_1, x_2,\ldots, x_l\}$ be an independent and identically distributed set of instances from a random variable X with spread bounded by R. Let $\bar{x}$ and s_x be the sample mean and standard deviation of the sampling distribution. Then for any $\delta \in [0,1]$, with probability at least $(1-\delta)$ then

$$\left| x_{l+1} - \bar{x} \right| \frac{R}{\sqrt{l}} \left(2 + \sqrt{2 + ln\frac{1}{\delta}} \right) = f(R,l,\delta)$$

where

1. f is an increasing function of R, which is the spread,

2. f is a decreasing function of l, the number of points, and

3. f is an increasing function of $1 - \delta$.

Using these tools the outliers can at least be identified. We start with a set of independently identically distributed data points with sample standard deviation, s_x, and mean $\bar{x}$. Upon the arrival of a new data point, the question to resolve is: "Is the new point an outlier?"

Basically the approach is to determine if the new data point lies outside a neighborhood centered at μ and with radius $max_{1 \leq i \leq l}(d_i)$ where $d_i = |x_i - \mu|$, then the new data point is an outlier. However we don't know μ and, therefore, don't know the d_i for $i \in [0,1]$. The solution is to utilize the sample mean and Hoeffding's inequality to estimate the needed threshold. Because we have symmetry and independently identically distributed data points, then:

$$P(max_{1 \leq i \leq l+1}\, d_i \neq max_{1 \leq i \leq l}\, d_i) = P(d_{l+1} > max_{1 \leq i \leq l}) \leq \frac{1}{l+1}$$

[8] Shawne-Taylor, J., & Cristianini, N. (2005). *Kernel Methods for Pattern Analysis,* Cambridge.

then

$$d_{l+1} = |x_{l+1} - \mu| \geq |x_{l+1} - \bar{x}| - |\bar{x} - \mu|$$

then for all i:

$$P\left(|x_{l+1} - \bar{x}|\right) > \left[\max_{1 \leq i \leq l} |x_i - \bar{x}| + 2f(R, \delta, 1)\right] < \frac{1}{l+1}.$$

In this section, all of the previous material applies to an univariate distribution. But in most applications a multivariate distribution is involved.

Now consider that we have a set of independently and identically distributed data points $\{x_1, x_2, \ldots, x_n\}$. Let the sample mean be $\bar{x}$ and we are given a new data point x_{l+1}. Now use $\bar{x}$ and the Hoeffding inequality to estimate a threshold, which enables the identification of outliers.

The multivariate normal distribution has the probability density function:

$$f(X = x) = e^{(x-\mu)^{TR}} \sum\nolimits^{-1} (x - \mu)$$ where $\sum$ is the $d \times d$ variance-covariance matrix. Let $X_{NR,ND}$ be the data matrix with NR rows and ND columns. The *Mahalanobis distance* between two points x and y in d dimensional space is:

$$\text{Mahal}(x, v) = (x - y) \sum\nolimits^{-1} (x - y)^{\text{Tr}}$$

and

$$\sum = \begin{bmatrix} \sigma_{11} & \sigma_{12} & \cdots & \sigma_{1d} \\ \sigma_{21} & \sigma_{22} & \cdots & \sigma_{2d} \\ \vdots & \vdots & \vdots & \vdots \\ \sigma_{d1} & \sigma_{d2} & \cdots & \sigma_{dd} \end{bmatrix}$$ is the variance-covariance matrix.

Completion of the following steps will construct the variance-covariance matrix:

Step 1: Subtract the mean vector from each row of the data matrix.

Step 2: Compute the dot product between the columns of the data matrix.

Step 3: Multiply the data matrix, after completion of Step 2, by $\frac{1}{NR-1}$.

The square of the Mahalanobis distance to the mean of the data set is a $\frac{1}{NR-1}$ chi-square distribution, with d degrees of freedom.

Using this information, the following algorithm can be employed for finding multivariate outliers:

Input: a $NR \times d$ data set X

Output: the candidate outliers

1. Calculate μ and $\sum$.

2. Let D be the $NR \times 1$ vector containing of the square of the Mahalanobis distance to μ.

3. Find the data points O in D whose value is greater than the critical value for χ_d^2 for $(1 - \alpha) = 0.975$.

4. Return O.

When applying this algorithm, the investigator must remember several compounding factors:

One outlier can skew the results of a study because the mean is very sensitive to extreme values.

The Mahalanobis distance itself is impacted by the outliers. Rousseeuw and Driessen[9] devised a method for making the statistical estimator less sensitive to the outliers, called the *Minimum Covariance Determinant, MCD*. Using a subset of n data points, the MCD minimizes the determinant of the variance-covariance matrix over all subsets of size n. Let

$R^* = \text{argmin} \{\det(\Sigma_R) | \text{R is a subset of the data matrix } D \text{ of size } n\}$. Then compute μ and Σ and finish by computing the Mahalanobis distance based not on D but on R^*. A very important result for the MCD is:

Let D *be the original* NR × d *data set and* R *be a size* n *subset of* D. *Compute* μ_R *and* Σ_R *based upon* R. *Compute the Mahalanobis distance of all points in* D *based upon* μ_R *and* Σ_R. *Sort the Mahalanobis distances and select* n *points with the smallest distance, label this set as* R2. *Then* $det(R2) \leq \det(R)$.

Therefore, when starting with a random configuration, the next generation will not increase the determinant. Then D-D plots, comparing the robust Mahalanobis to the full Mahalanobis will detect some candidate outliers.

---

[9] Rousseeuw, P. J., & Driessen, K. V. (1999). A fast algorithm for minimum covariance determinant estimator. *Technometrics*, 41:212-223.

This section has presented a very brief introduction on statistical algorithms for detecting outliers. Neighborhoods are part of the foundation for the statistical methods. A natural outgrowth of these approaches leads to using clustering methods to find clusters with few data points that are expected to contain observations that are significantly different from the majority of the data points.

10.4 OUTLIER DETECTION BY CLUSTERING

An algorithm for outlier detection is to:

Step 1: run a hierarchical clustering method on the data set.

Step 2: select the data points, or rows of the data set matrix, that are allocated to small clusters.

Step 3: manually correct the data points from noise, etc.

Several parameters need to be specified. The level for cutting the dendrogram needs to be specified. If the level number is small, then the outliers can end up being members of clusters based upon normal observations. If the number of clusters is too large, then the end result might be selecting several clusters containing only normal members. Choice of hierarchical clustering needs to be addressed. Another parameter to consider is the choice of distance function. Before selecting the small clusters, the investigator could consider an extra step in the algorithm for optimization, using a K-means type clustering. But the K-means algorithm is sensitive to noise and outliers (Laan, Pollard, and Bryan).[10]

Clustering-based approaches are capable of being used incrementally, or after determining the clusters then new data points can be inserted into the system and tested for outliers. Normal points tend to belong to large densely filled clusters, whereas outliers do not belong to any cluster or to a small sparsely filled cluster. This is one of the possible environmental constraints that is in place when using cluster analysis for outlier detection. Wu and Zhang[11] used normal data to generate clusters for representing normal

---

[10] Laan, Pollard, M. K., & Bryan, J. (2003). A new partitioning around medoids algorithms. *Journal of Statistical Computation and Simulation*, (73), No. 8, 575-584.

[11] Wu, N., & Zhang, J. (2003). Factor analysis based anomaly detection. In *Proceedings of IEEE Workshop on Information Assurance*. United States Military Academy, West Point, NY, USA.

modes of behavior of the training data. Then any new data instance should be posted to one of the existing clusters if it is a normal data; otherwise, it is identified as an outlier. He, Xiaofei, and Shengchun[12] calculated a measure called the semantic outlier factor which is high if the class label for an object in a cluster is different from the majority of the class labels in that cluster. A nearest neighbor analysis is widely used for outlier detection. Normal points are, by definition, points having normal closely related neighbors and outliers are points which are far from other points. Knorr and Ng[13] proposed generalizations of outliers based on a variety of distributions. These generalizations are similar to those proposed by Kollios, Gunopulos, Koudas, and Berchtold[14] and Ramaswamy, Rastogi, and Shim.[15] Kollios, Gunopulos, Koudas, and Berchtold give the following definition for outliers:

An object O in a data set T is a *DB(k,D) outlier* if at most k objects in T lie a distance at most D from O. Ramaswamy, Rastogi, and Shim's definition is:

Outliers are the top n data elements whose distance to the kth nearest neighbor is greatest. These authors recommend the application of a simple nested loop algorithm:

For each object $O \in T$, compute the distance to each object $(q \neq O) \in T$ until $(k + 1)$ neighbors are found with distance less than or equal to D.

If $|Neighbor(O)| \leq k$, then report O as a *DB(k,D)* outlier.

Zhang and Wang[16] defined the outlier score of a data point to be the sum of its distances from its k nearest neighbors and additionally proposed a post processing of the outliers to identify the subspaces in which they exhibited outlying behavior.

---

[12] He, Z., Xiaofei, X., & Shengchun, D. (2002). Squeezer: An efficient algorithm for clustering categorical data. *Journal of Computer Science and Technology,* 17, 5, 611-624.

[13] Knorr, E. M., & Ng, R. T. (1999). Finding intentional knowledge of distance-based outliers. In *Proceedings of 25th International Conference on Very Large Data Bases,* Morgan Kaufmann, 211-222.

[14] Kollios, G., Gunopulos, D., Koudas, N., & Berchtold, W. (2003). Efficient biased sampling for approximate clustering and outlier detection in large datasets. *IEEE Transactions on Knowledge and Data Engineering.*

[15] Ramaswamy, S., Tastogi, R., & Shim, K. (2000). Efficient algorithms for mining outliers from large data sets. In *SIGMOD Conference,* 427-438.

[16] Zhang, J., & Wang, H. (2006). Detecting outlying subspaces for high-dimensional data: the new task, algorithms, and performance. *Knowledge and Information Systems* 10, 3, 333-355.

All of the techniques discussed so far locate global outliers and perform poorly if varying degrees of density are present among the data.

Breunig, Kriegel, No, and Sander[17] use LOF for computing the density of regions in the data and declare the instances in low dense regions as outliers. Although local outlier detection methods overcome the variation in density issues in the data, they lead to a computational explosion. The major problem with nearest neighbor methods is that the computational complexity required to compute the distances between every point is $O(n^2)$. However, Bay and Schwabacher[18] proposed a *linear time algorithm*, rather than the common quadratic time algorithm, using randomization and pruning for detecting outliers:

1. Randomize the data.

2. Partition the data matrix into blocks.

3. Compare each point in the block to every point in the data matrix.

4. Keep track of the Top n outliers, Top n, and the weakest outlier, the point in Top n which has the smallest k nearest neighbor, during block processing.

5. Prune points as soon as they become nonoutliers.

6. As more blocks are processed, the weakest score keeps increasing and more points get pruned sooner.

One of the most commonly used algorithms, DBSCAN (Density-Based Spatial Clustering of Applications with Noise), captures outliers. DBSCAN, developed by Ester, Kriegel, Sander, and Xu,[19] determines a number of clusters based upon the density distribution of neighboring points. Clusters keep growing during execution of the DBSCAN algorithm, until the density, or number of points, in the neighborhood exceeds some threshold.

[17] Breuunig, M. M., Kriegel, H. P., No, R. T., & Sanders, J. (2000). LOF: identifying density-based local identifiers. In *Proceedings of 2000 ACM SIGMOD International Conference on Management of Data,* ACM Press, 93-104.

[18] Bay, S. D., & Schwabacher, M. (2003). Mining distance-based outliers in near linear time with randomization and a simple pruning rule. In *Proceedings of Ninth ACM SIGKDD International Conference on Knowledge Discovery and Data Mining,* Washington, D. C., USA, 29-38.

[19] Ester, M., Kriegel, H-P, Sander, J., & Xu, X. (1996). A density-based algorithm for discovering clustering large spatial databases with noise. *Proceedings 2nd International Conference on Knowledge Discovery and Data Mining.*

Clusters located by DBSCAN are sets of "density connected" points. DBSCAN clusters are composed of four types of points: the core point (or center of the neighborhood), interior points, border points, and noise points. The operational definition for these points is dependent on the definition of density reachable.

Definition: **Eps** is the distance from the core point.

Definition: **MinPts** is the specified minimal number of points in a neighborhood to be able to have the neighborhood labeled as a cluster.

Definition: The **Eps Neighborhood** is the neighborhood within a radius of eps from the specified point.

Definition: If the eps neighborhood of a point contains at least MinPts then the point is called a **core point.**

Definition: A point q is **directly density-reachable** from a point p if it is not a greater distance than a specified distance ε, or the eps parameter for the DBSCAN algorithm, and if p is surrounded by MinPts or more points such that p and q are part of a cluster.

Definition: A **border point** has fewer than MinPts within eps, but is in the neighborhood of a core point.

Definition: A **noise point** is any point that is not a core point nor a border point.

Figure 10.1 illustrates a core point, a 1-neighborhood of q, and a noise point.

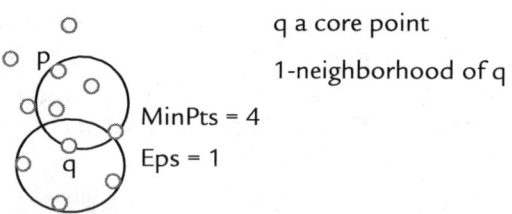

q a core point

1-neighborhood of q

MinPts = 4

Eps = 1

FIGURE 10.1 A Core Point q Using a 1-Neighborhood and a Noise Point p.

Definition: A point p is **density-reachable** from a point q given eps and MinPts if there a sequence $p_1, p_2, ..., p_m$ where $p_1 = p$ and $p_m = q$ with each p_{i+1} being directly density reachable from p_i. Figure 10.2 illustrates a pair of points p and q where q is density-reachable from p.

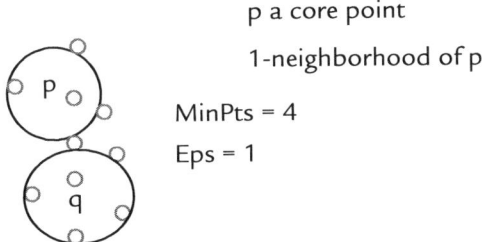

p a core point

1-neighborhood of p

MinPts = 4

Eps = 1

FIGURE 10.2 Point p is Density-Reachable from Point q.

Note that the relationship of being density-reachable is not symmetric, as illustrated in Figure 10.3.

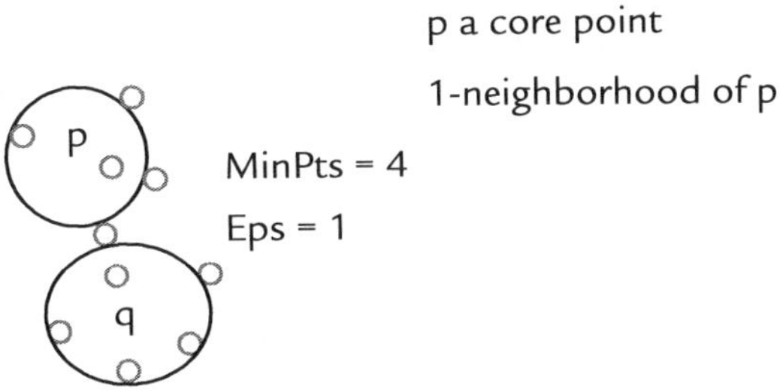

p a core point

1-neighborhood of p

MinPts = 4

Eps = 1

FIGURE 10.3 Point p is not Density-Reachable From Point q.

Definition: Two points p and q are **density-connected** if there is a point r where p and r as well as r and q are density-reachable.

In Figure 10.4 points p and q are density-connected.

MinPts = 3

Eps Eps specified

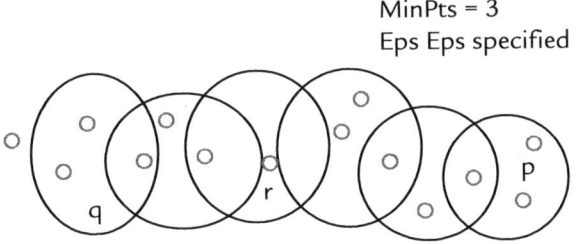

FIGURE 10.4 Density-Connected Points p and q.

Definition: A **cluster**, C, is a subset of the points in the database where:

- All points in C are mutually density-connected, and

- If a point p is density-connected to any point in C then p is a member of C.

The following pseudo-code for the DBSCAN algorithm has Eps and MinPts as parameters;

DBSAN Algorithm

```
Step 1: Input Eps and MinPts plus the database, D.
Step 2: C = 0 {label for cluster 0}
Step 3: Repeat for each unvisited p in D
        { mark p as visited
        N ← neighbors(p,Eps)
        IF size(N) < MinPts,
                THEN Mark p as NOISE
        ELSE
                {C = C + 1
                Expandcluster(p, N, C, Eps, MinPts)}
}
Expandcluster(p, N, C, Eps, MinPts)
        Post p to C
        Repeat for each point q in N
        {IF q is not visited
                THEN mark q as visited
        N' = neighbors(q, Eps)
        IF size(N') ≥ MinPts
                THEN N = N∪N'
        IF q is not a member of any cluster
                THEN Post q to C}
```

Notice that unlike k-means DBSCAN does not require the end user to know the number of clusters ahead of time. Besides finding outliers, DBSCAN can locate arbitrarily shaped clusters. On the negative side, DBSCAN does not work well if the data sets in the database have varying densities or a bad choice is made for the distance measure used by the function neighbors (p, Eps).

10.5 FUZZY CLUSTERING OUTLIER DETECTION

Belal, Al-Zoubi, Al-Dahoud, and Yahya[20] use the Fuzzy C-means clustering algorithms for outlier detection. The advantages for this choice include: implementation ease, applicability to multidimensional data, and the ability to model uncertainty within the data. In phase one, a Fuzzy C-means

---

[20] Belal, M., Al-Zoubi, Al-Dahoud, & Yahya, A. A. (May, 2010). New Outlier Detection Method Based on Fuzzy Clustering. *WSEAS Transactions on Information Science and Applications,* Issue 5, (7): 661-690.

clustering is run on the data set in order to produce an objective function. This objective function represents the Euclidean distance between cluster centroids. Data points belonging to the resultant clusters are multiplied by the membership values of each cluster produced by the Fuzzy C-means clustering. During phase two, small clusters are then determined and considered as outlier clusters. Small clusters are clusters with fewer data points than half the average number of data points in the c-clusters. This is a density-based outlier method, not a distance-based method.

Basically, this method emphasis is on the assumption that removing a data point will cause a decrease in the objective function, the total sum of squares of distances between the cluster centroids and the points that are members of the clusters. If the decrease is greater than a specified threshold, then the data point is considered to be an outlier.

10.6 SUMMARY

- Given a set of identically independently distributed points $\{x_1, \ldots x_n\}$ with sample mean s and population mean μ. A new point x_{n+1} arrives. Is x_{n+1} an outlier? Answer: Let B be a ball centered at μ of radius equal to the maximum distance between data points in the ball and the ball's centroid. If the new data point lies outside the ball then declare it as an outlier.

- It is well known that both the mean and standard deviation are extremely sensitive to outliers.

- "Robustification" means making the statistical estimator less sensitive to outliers.

- The weakness of statistical outlier methods is their sensitivity to the mean of the data sets.

- Distance-based outlier detection methods are very sensitive to varying degrees in cluster densities in a clustering.

10.7 EXERCISES

1. Distinguish between distance-based and density-based outliers.

2. Find the outlier(s) for the data set {25, 73, 34, 85, 95, 38, 79, 56, 38, 67, 74} using the local outlier factor method.

3. Find the outlier(s) for the data set {25, 73, 34, 85, 95, 38, 79, 56, 38, 67, 74} using the relative density approach.

4. Find the outlier(s) for the data set {25, 73, 34, 85, 95, 38, 79, 56, 38, 67, 74} using Chebychev's inequality.

5. Find the outlier(s) for the data set {25, 73, 34, 85, 95, 38, 79, 56, 38, 67, 74} using the Empirical Rule.

6. Find the outlier(s) for the data set {25, 73, 34, 85, 95, 38, 79, 56, 38, 67, 74} using the Hoeffding Like Weak Inequality.

7. Filzmoser[21] simulates a data set in two dimensions in order to simplify the graphical visualization. Eighty-five data points follow a bivariate standard normal distribution. Multivariate outliers are introduced by 15 points coming from a bivariate normal distribution with mean $(2; 2)T$ and covariance matrix diag(1=10; 1=10).

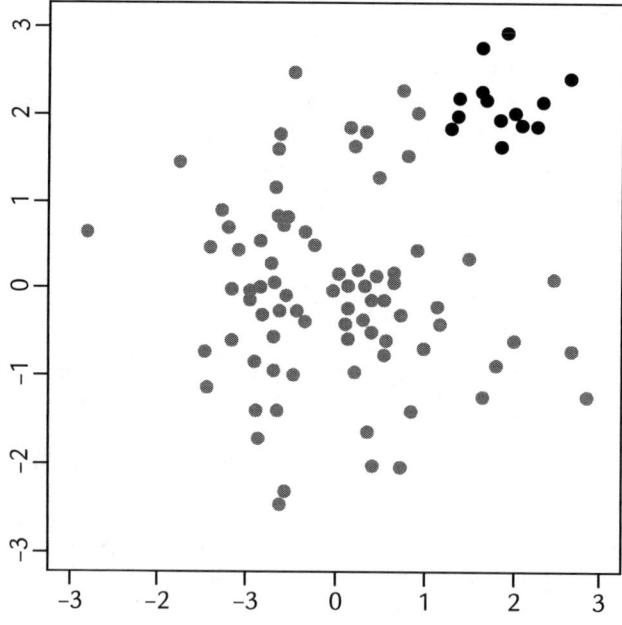

FIGURE 10.5 Filzmoser's Simulated Data Set.

Describe how the Mahalanobis distance can be used to find the outliers in this data set.

---

[21] Filzmoser, P. A multivariate outlier detection, e-mail: P.Filzmoser@tuwien.ac.at

8. Describe how clustering methods could be used to find the outliers in the data set for Exercise 7.

9. Describe how fuzzy clustering methods could be used to find the outliers in the data set for Exercise 7. Which method is preferable, the crisp clustering or the fuzzy clustering method? Why?

10. Use the Neymann-Scott program listing and the linear time algorithm by Bay and Schwabacher to implement a program to run an outlier detection for 2 dimensional data sets with 3 to 5 features.

11. Defend the statement: "It is well known that both the mean and standard deviation are extremely sensitive to outliers."

12. Defend the statement: "The weakness of statistical outlier methods is their sensitivity to the mean of the data sets."

13. Defend the statement: "Distance-based outlier detection methods are very sensitive to varying degrees in cluster densities in a clustering."

14. Use the Neymann-Scott program listing and the linear time algorithm by Bay and Schwabacher to implement a program to run an outlier detection for 2 dimensional data sets with 3 to 5 features and using an optimizing clustering approach.

15. Consider the following data points:

Point	X	Y
1	1	1
2	2	2
3	3	3
4	4	4
5	4	5
6	4	6
7	5	5
8	6	6
9	7	7
10	8	8
11	7	6
12	6	7
13	6	8
14	8	6
15	9	10
16	3	7
17	7	3
18	6	4
19	4	6
20	4	3
21	3	2

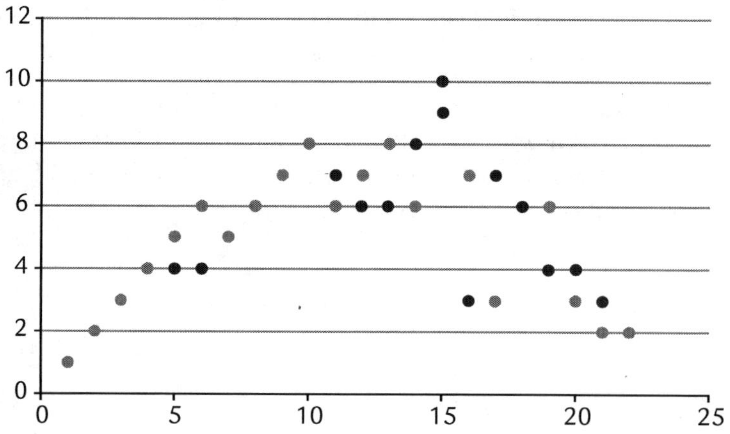

FIGURE 10.6 Hypothetical Data Set.

(a) Illustrate points in the data set that are directly density reachable.

(b) Illustrate points in the data set that are density connected but not density reachable.

16. Use, by hand, the DBSCAN algorithm to obtain a clustering for the data set in Exercise 15.

17. Implement and run the DBSCAN algorithm, in a programming language of your choice, on the data set for Exercise 12.

CHAPTER 11

MODEL-BASED CLUSTERING

In This Chapter

11.1 INTRODUCTION

Model-based clustering assumes that the data is based upon a mixture of probability distributions given that the cluster memberships of the data set are not known. The objective for model-based clustering is to estimate the parameters of the cluster distributions by maximizing the likelihood function of the mixture density with respect to the observed data. In general model-based clusterings are either statistical or AI algorithms. Neural networks will be discussed in Chapter 12.

An example of model-based clustering is COBWEB, a conceptual clustering system that organizes data to maximize inference abilities (Fisher).[1]

---

[1] Fisher, D. (1987). Improving Inference Through Conceptual Clustering. *AAAI-87 Proceedings.*

COBWEB was developed as a model to simulate how humans incrementally form concepts. The concept hierarchy is represented by a generalized tree structure where each new instance is represented as an ordered set of attribute-value pairs. Each attribute can be assigned only one value per instance. During tree construction, the tree is recursively navigated using a form of the hill-climbing search rather than a depth-first or breadth-first search. COBWEB stores the probability of the concept's occurrence at each concept node, as well as information about every attribute observed in the instances that are covered by the concept. As COBWEB navigates the tree structure, it selects which tree operations to apply using a category utility, which gives a high score to partitions that maximize intra-class similarity and inter-class similarity.

11.2 COBWEB: A STATISTICAL AND AI APPROACH

Consider the construction of a hierarchy of abstract classes for the following set of traffic signs:

Sign number	Sign	Color	Shape	Type
1	Yield	Yellow	Triangle	Warning
2	Stop	Red	Octagonal	Regulatory
3	Railroad crossing	Yellow	Round	Warning
4	Bicycle route	Green	Rectangle	Guide
5	Do not enter	Red	Round	Regulatory
6	Intersection	Yellow	Diamond	Warning
7	School zone	Yellow	Pentagon	Warning
8	National park	Brown	Trapezoid	Guide

TABLE 11.1 Traffic Signs Data Set.

The series of traffic sign instances are input to the COBWEB algorithm and then stored as a concept node, or as a knowledge representation. The concept hierarchy is a tree where each node in the tree describes a concept. The first instance, a yield sign, is described as an ordered set of attribute-value pairs as illustrated in Table 11.2:

Color	Yellow
Shape	Triangle
Type	Warning

TABLE 11.2 Instance of a Yield Sign.

Each time an instance is added to the hierarchy, knowledge is added by changing information within the concept nodes or possibly by changing the overall structure of the hierarchy. The first instance establishes the root of the tree as depicted by Table 11.1:

First Instance:
Yield sign

Color	Yellow
Shape	Triangle
Type	Warning

Hierarchy:

$P(C_0)$ =1.00		$P(V|C)$
Color	Yellow	1.00
Shape	Triangle	1.00
Type	Warning	1.00

FIGURE 11.1 Concept Hierarchy After the First Instance of a Traffic Sign.

COBWEB stores the conditional probability of each given attribute value, given the membership in the class covered by the concept. Additionally COBWEB stores the probability of the concept's occurrence at each concept node. The nodes in the hierarchy are numbered as C_k for k = 0 to 8. Terminal nodes in the hierarchy describe single instances. The top of each node is the probability of occurrence for that specific class, or the parent for a specific class.

As instances are input to COBWEB, four operations can be applied to the hierarchy for incorporating new knowledge. These operations are applied locally to the sub-tree composed of the last concept for which an instance was classified and to the children of this node. An evaluation function, called the category utility, is used by COBWEB to select which operation to apply. This selection is to choose the maximum value of the category utility value.

The four operations are:

1. To *incorporate*: when an instance fits into an existing concept. The instance is integrated into one of the child nodes. If the child node is not a singleton, then the conditional probabilities are updated in the concept node and each of the attribute values is updated. If the node being incorporated is a singleton then the incorporating node is added as a new downward leaf node.

2. To *create new disjunct*: when an instance has very different characteristics from any existing concept at the current concept level. The instance is inserted as a category by itself or a sibling of the existing concept nodes.

3. To *merge*: occurs when the hierarchy is overly branched. Two classes are combined to provide a good concept to which to classify the incoming instance.

4. To *split*: when the hierarchy contains a node that is too general. The general node is broken into well-defined classes to create a good match for the incoming node by removing the current node and replacing it with its children.

To make this choice, the system must be able to evaluate alternative classifications and apply the operator that produces the best hierarchy. COBWEB uses the *category utility* to score these alternatives. Different classifications of a new instance result in a number of different hierarchies of all the instances into classes. The category utility function gives a high score to partitions which maximize similarity among class members, intra-class similarities, and differences between members of different classes, inter-class differences.

In effect, the category utility trades off the predictiveness of each attribute value, the probability of an instance's membership in a class, given its attribute value, and the *predictivity* of the value, the probability of the value, given that an instance is a member of a class. To summarize, predictive values are those most nearly unique to a certain class and therefore indicative of it. The evaluation function favors classes with many predictive attribute values because these maximize inter-class differences. Predictable values are those that many members share and therefore are easy to guess accurately. The evaluation function favors classes with many predictable values because these maximize intra-class similarities. Because attributes are often not both predictable and predictive, category utility trades off the two, maximizing each as much as possible. The category utility equation can be summarized as

$$\frac{X - Y}{K},$$

where X is the expected number of attribute values that can be correctly guessed, given the K categories, and Y is the expected number of attribute values that can be correctly guessed without any category knowledge. Dividing by K, the total number of classes, normalizes for partitions with differing numbers of classes. In the expanded equation, the X term is

$$\sum\nolimits_{k=1}^{K} P(C_k) \sum\nolimits_{i=1}^{I} \sum\nolimits_{j=1}^{J} P(A_i = V_{ij} \lfloor C_k)^2$$

summing across K classes, I attributes, and J values. $P(C_k)$ is the probability of occurrence of a particular class C_k and $P(A_i = V_{ij} \lfloor C_k)$ is the conditional probability of a particular value V_{ij} given membership in the class. The Y term expands to

$$\sum\nolimits_{i=1}^{I} \sum\nolimits_{j=1}^{J} P(A_i = V_{ij})^2$$

where $P(A_i = V_{ij})$ is the probability of a particular value at the parent of the node classes being considered; that is, the probability across all classes without category knowledge. The complete equation is:

$$\frac{\sum\nolimits_{k=1}^{K} P(C_k) \sum\nolimits_{i=1}^{I} \sum\nolimits_{j=1}^{j} [P((A_i = V_{ij}) \lfloor C_k)^2 - P(A_i = V_{ij})^2]}{K}.$$

For information on the derivation of this equation, refer to Gluck and Corter,[2] which gives a two-class version of the equation, and Fisher,[3] which gives this multi-class form.

Figure 11.3 depicts the hierarchy generated by creating a new adjunct operation after the second instance is input.

The following navigation path is followed to generate Figure 11.2. First an attempt is made to insert or incorporate at the root node. Because this is a distinct sign separate from the root node, then a new concept node, based upon any of the three attributes needs to be considered. For all three attributes: $X = (0.5)(1^2 + 1^2 + 1^2) + (0.5)(1^2 + 1^2 + 1^2) = 3$; $Y = 3((0.5)^2 + (0.5)^2) = 1.5$;

[2] Gluck, M., & Corter, J. (1965). Information, uncertainty and the utility of categories. *Proceedings of the Seventh Annual Conference of the Cognitive Science Society.* Irvine, CA: Lawrence Erlbaum, 283-287.

[3] Fisher, D. (1987). *Knowledge acquisition via incremental conceptual clustering,* Doctoral dissertation, Department of Information and Computer Science, University of California, Irvine.

Second Instance:
Stop sign

Color	Red
Shape	Octagonal
Type	Regulatory

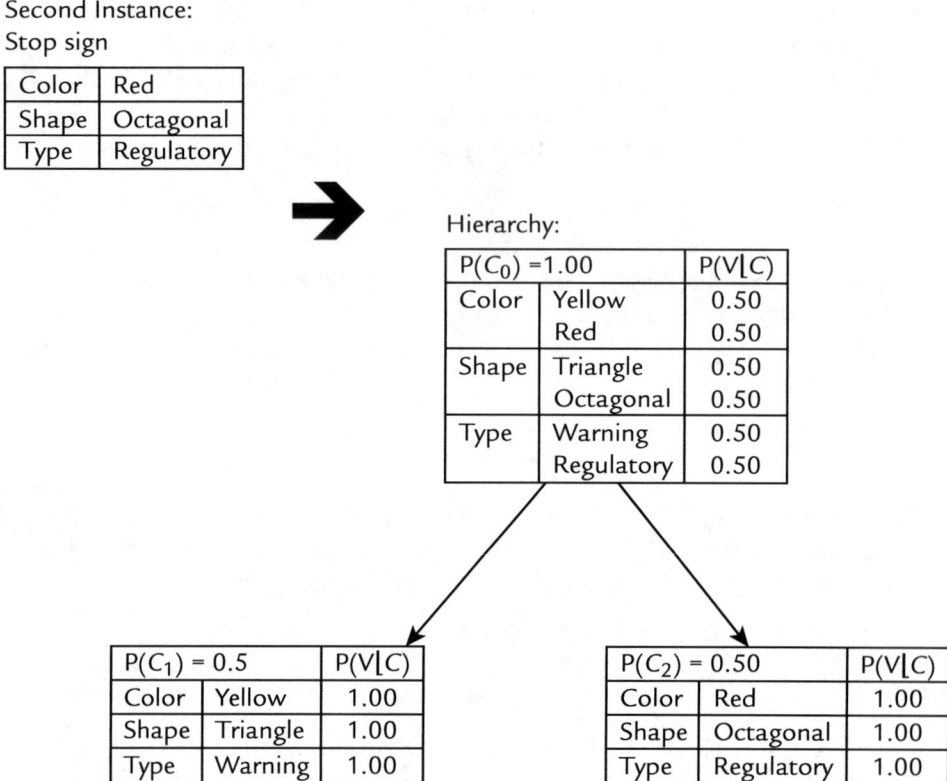

Hierarchy:

$P(C_0)$ =1.00		$P(V|C)$
Color	Yellow	0.50
	Red	0.50
Shape	Triangle	0.50
	Octagonal	0.50
Type	Warning	0.50
	Regulatory	0.50

$P(C_1)$ = 0.5		$P(V|C)$
Color	Yellow	1.00
Shape	Triangle	1.00
Type	Warning	1.00

$P(C_2)$ = 0.50		$P(V|C)$
Color	Red	1.00
Shape	Octagonal	1.00
Type	Regulatory	1.00

FIGURE 11.2 Concept Hierarchy After the Second Instance of a Traffic Sign.

and the category utility value is $\dfrac{3-1.5}{2}$ = 0.75. Therefore any of the attributes can be chosen for the root node linked list header pointer. Each concept node has three link list headers: color list header, shape list header, and the type linked list header. For class C_0 node in Figure 11.2 both the shape list header and the type list header are set to null, and in this underlying data structure for the hierarchy tree, the color list header points to a FIFO linked list containing the first and second instances.

When inputting the third instance, the railroad crossing sign, first the root directs the instance to the class C_1 node. Then either the third instance should be entered into the color link list of the root node, the C_0 node, using the incorporate operation or merge operation applied to the C_1 node. For the incorporate operation, the value of $X = (0.33)(1^2 + 0.5^2 + 0.5^2 + 1^2) + (0.33)(1^2 + 1^2 + 1^2) + (0.33)(1^2 + 1^2 + 1^2) = 2.81$; $Y = 1^2 + 0.5^2 + 0.5^2 + 1^2 = 2.5$; and the category utility value is 0.10. For the merge operation the value of $X = (0.5)(1^2 + 1^2 + 1^2) + (0.5)(1^2 + 1^2 + 1^2) = 3$; $Y = 1^2 + 0.5^2 + 0.5^2 + 1^2 = 2.5$; and the category utility value is 0.17. Therefore, a merge of class C_1 node generates Figure 11.3

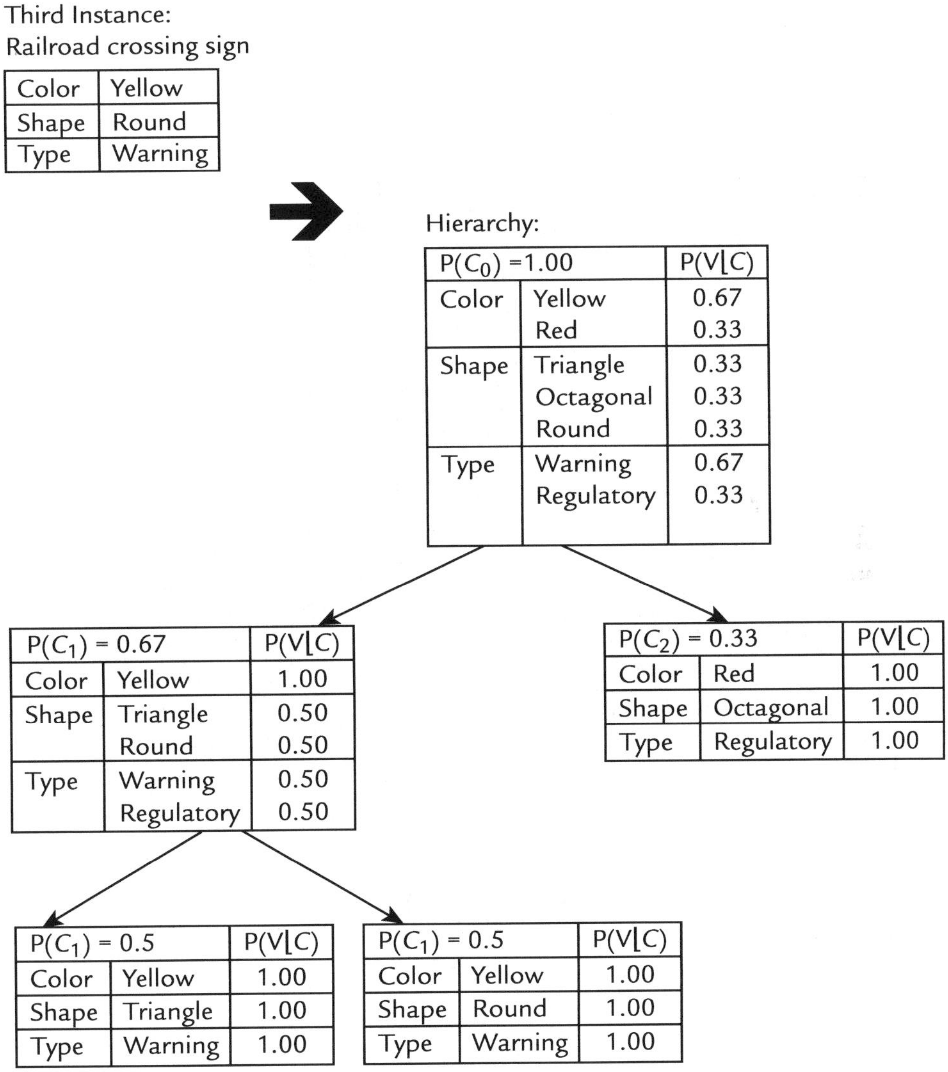

Third Instance:
Railroad crossing sign

Color	Yellow
Shape	Round
Type	Warning

Hierarchy:

$P(C_0) = 1.00$		$P(V \lfloor C)$
Color	Yellow	0.67
	Red	0.33
Shape	Triangle	0.33
	Octagonal	0.33
	Round	0.33
Type	Warning	0.67
	Regulatory	0.33

$P(C_1) = 0.67$		$P(V \lfloor C)$
Color	Yellow	1.00
Shape	Triangle	0.50
	Round	0.50
Type	Warning	0.50
	Regulatory	0.50

$P(C_2) = 0.33$		$P(V \lfloor C)$
Color	Red	1.00
Shape	Octagonal	1.00
Type	Regulatory	1.00

$P(C_1) = 0.5$		$P(V \lfloor C)$
Color	Yellow	1.00
Shape	Triangle	1.00
Type	Warning	1.00

$P(C_1) = 0.5$		$P(V \lfloor C)$
Color	Yellow	1.00
Shape	Round	1.00
Type	Warning	1.00

FIGURE 11.3 Concept Hierarchy After the Third Instance of a Traffic Sign.

The fourth instance, a Bicycle route sign, is placed at the first node in the root node's linked list, therefore the concept hierarchy tree as illustrated in Figure 11.4. Note that this hierarchy was generated by applying an incorporation to a singleton while extending downward operation.

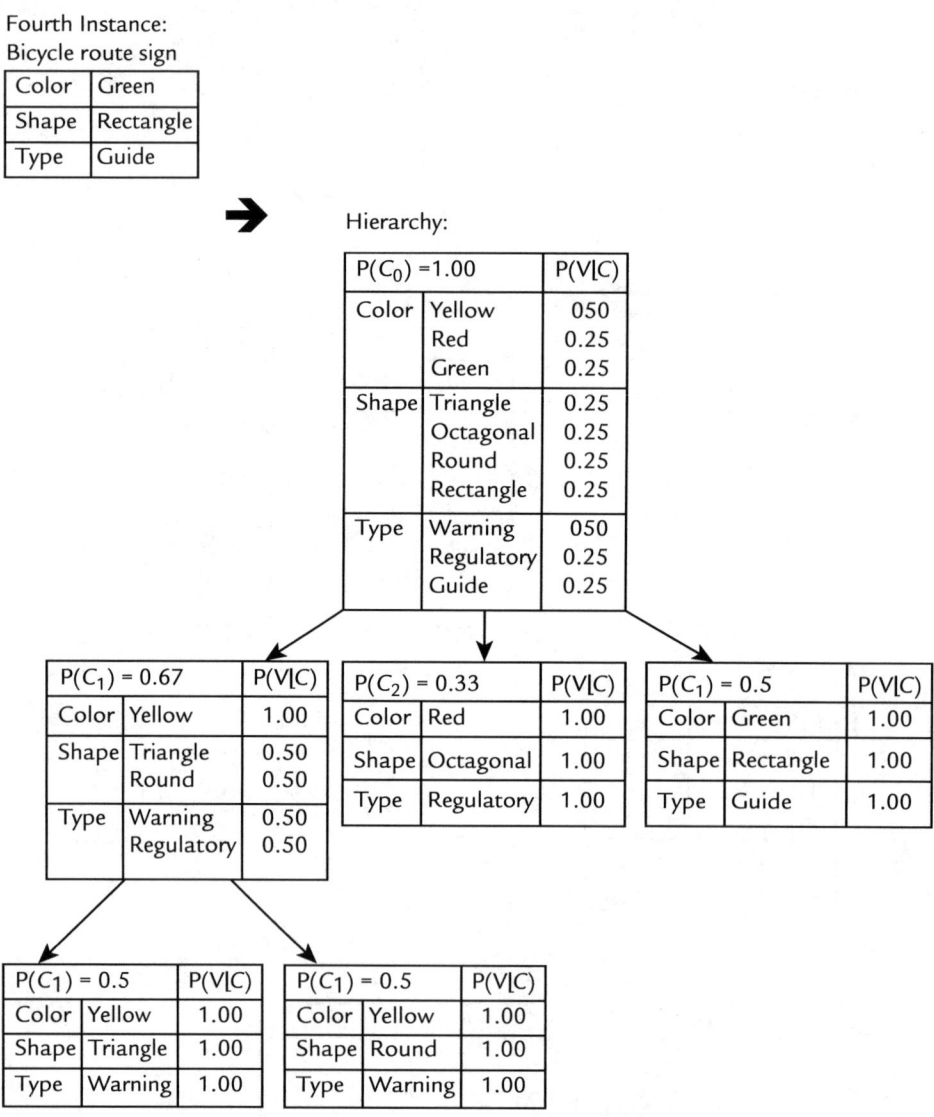

Fourth Instance:
Bicycle route sign

Color	Green
Shape	Rectangle
Type	Guide

➡

Hierarchy:

$P(C_0) = 1.00$		$P(V|C)$
Color	Yellow	050
	Red	0.25
	Green	0.25
Shape	Triangle	0.25
	Octagonal	0.25
	Round	0.25
	Rectangle	0.25
Type	Warning	050
	Regulatory	0.25
	Guide	0.25

$P(C_1) = 0.67$		$P(V|C)$
Color	Yellow	1.00
Shape	Triangle	0.50
	Round	0.50
Type	Warning	0.50
	Regulatory	0.50

$P(C_2) = 0.33$		$P(V|C)$
Color	Red	1.00
Shape	Octagonal	1.00
Type	Regulatory	1.00

$P(C_1) = 0.5$		$P(V|C)$
Color	Green	1.00
Shape	Rectangle	1.00
Type	Guide	1.00

$P(C_1) = 0.5$		$P(V|C)$
Color	Yellow	1.00
Shape	Triangle	1.00
Type	Warning	1.00

$P(C_1) = 0.5$		$P(V|C)$
Color	Yellow	1.00
Shape	Round	1.00
Type	Warning	1.00

FIGURE 11.4 Concept Hierarchy After the Fourth Instance of a Traffic Sign.

After the completion of the fifth instance, using a merge operation, the hierarchy is depicted as in Figure 11.5:

Fifth Instance:
Do Not Enter sign

Color	Red
Shape	Round
Type	Regulatory

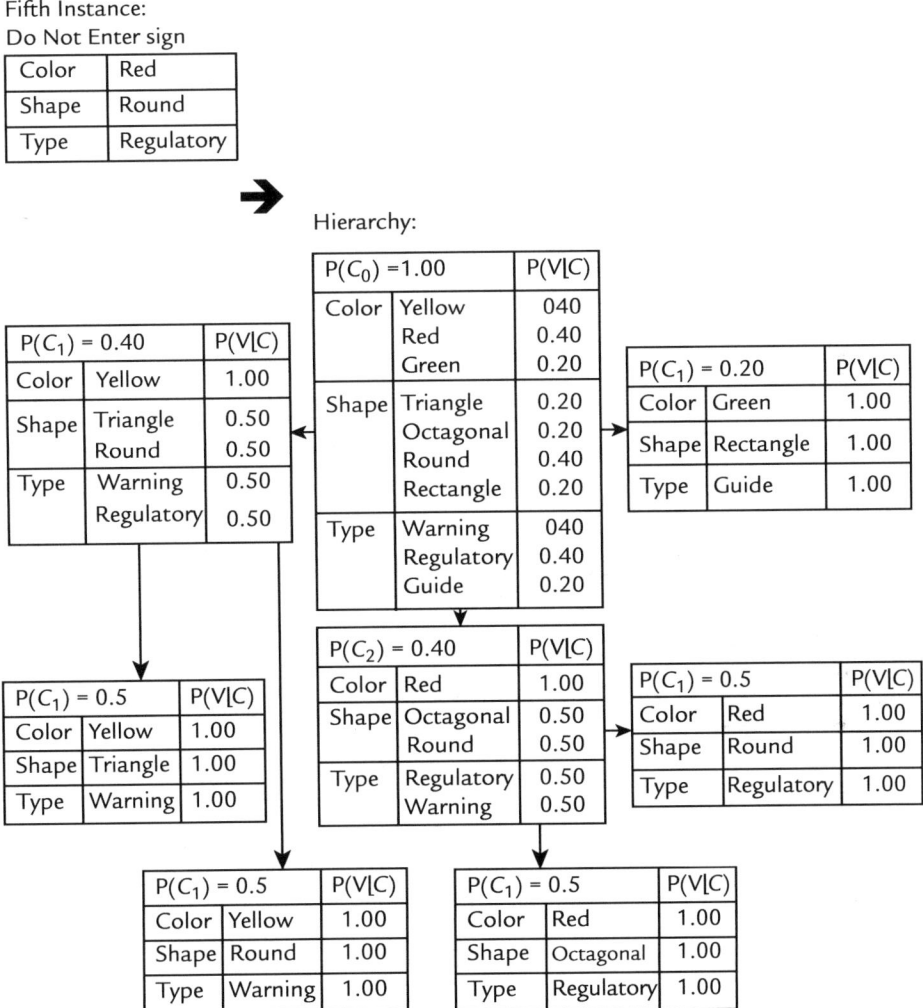

Hierarchy:

P(C_0) =1.00		P(V|C)
Color	Yellow	040
	Red	0.40
	Green	0.20
Shape	Triangle	0.20
	Octagonal	0.20
	Round	0.40
	Rectangle	0.20
Type	Warning	040
	Regulatory	0.40
	Guide	0.20

P(C_1) = 0.40		P(V|C)
Color	Yellow	1.00
Shape	Triangle	0.50
	Round	0.50
Type	Warning	0.50
	Regulatory	0.50

P(C_1) = 0.20		P(V|C)
Color	Green	1.00
Shape	Rectangle	1.00
Type	Guide	1.00

P(C_2) = 0.40		P(V|C)
Color	Red	1.00
Shape	Octagonal	0.50
	Round	0.50
Type	Regulatory	0.50
	Warning	0.50

P(C_1) = 0.5		P(V|C)
Color	Yellow	1.00
Shape	Triangle	1.00
Type	Warning	1.00

P(C_1) = 0.5		P(V|C)
Color	Red	1.00
Shape	Round	1.00
Type	Regulatory	1.00

P(C_1) = 0.5		P(V|C)
Color	Yellow	1.00
Shape	Round	1.00
Type	Warning	1.00

P(C_1) = 0.5		P(V|C)
Color	Red	1.00
Shape	Octagonal	1.00
Type	Regulatory	1.00

FIGURE 11.5 Concept Hierarchy After the Fifth Instance Added.

The sixth instance, an intersection sign, employs a new disjoint and the category utility function value is calculated for both the shape and type attributes, resulting in Figure 11.6.

Fifth Instance:
Do Not Enter sign

Color	Red
Shape	Round
Type	Regulatory

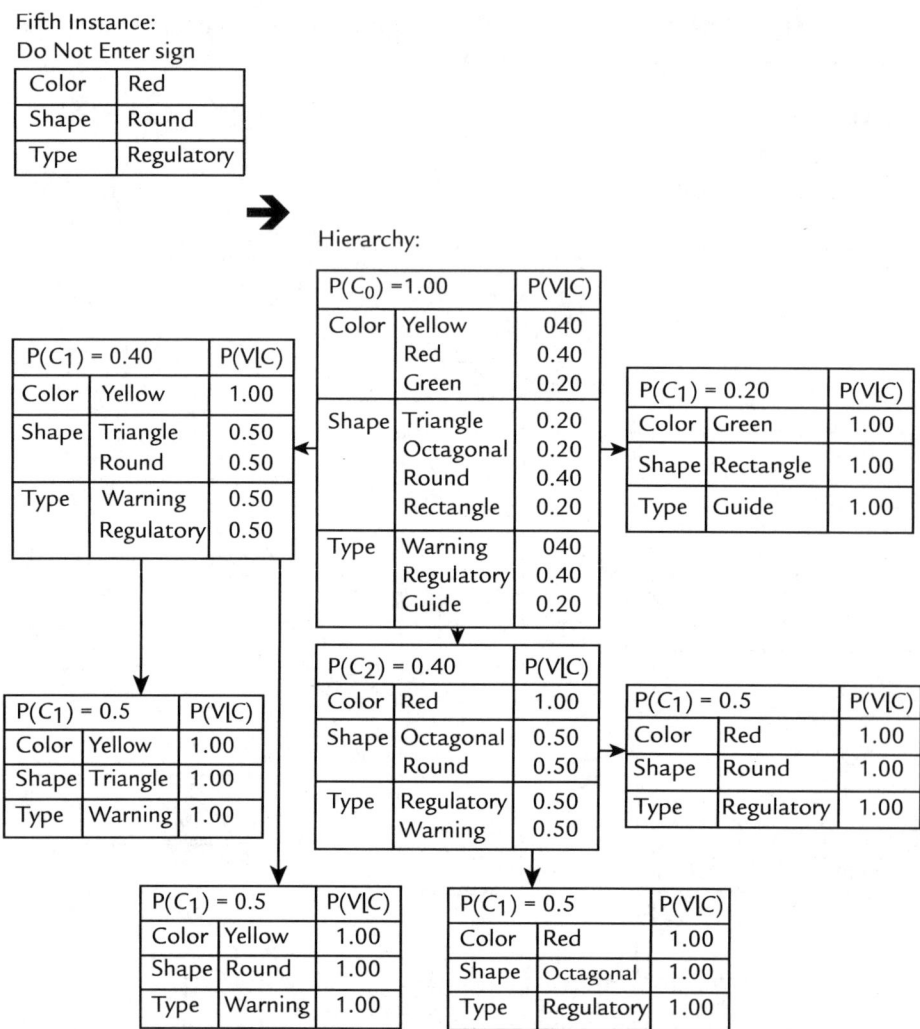

Hierarchy:

$P(C_0)$ =1.00		$P(V|C)$
Color	Yellow	040
	Red	0.40
	Green	0.20
Shape	Triangle	0.20
	Octagonal	0.20
	Round	0.40
	Rectangle	0.20
Type	Warning	040
	Regulatory	0.40
	Guide	0.20

$P(C_1)$ = 0.40		$P(V|C)$
Color	Yellow	1.00
Shape	Triangle	0.50
	Round	0.50
Type	Warning	0.50
	Regulatory	0.50

$P(C_1)$ = 0.20		$P(V|C)$
Color	Green	1.00
Shape	Rectangle	1.00
Type	Guide	1.00

$P(C_2)$ = 0.40		$P(V|C)$
Color	Red	1.00
Shape	Octagonal	0.50
	Round	0.50
Type	Regulatory	0.50
	Warning	0.50

$P(C_1)$ = 0.5		$P(V|C)$
Color	Yellow	1.00
Shape	Triangle	1.00
Type	Warning	1.00

$P(C_1)$ = 0.5		$P(V|C)$
Color	Red	1.00
Shape	Round	1.00
Type	Regulatory	1.00

$P(C_1)$ = 0.5		$P(V|C)$
Color	Yellow	1.00
Shape	Round	1.00
Type	Warning	1.00

$P(C_1)$ = 0.5		$P(V|C)$
Color	Red	1.00
Shape	Octagonal	1.00
Type	Regulatory	1.00

FIGURE 11.6 Concept Hierarchy After the Fifth Instance Added.

11.3 MIXTURE MODEL FOR CLUSTERING

COBWEB applies this version of category utility when instances have nominal attributes. It cannot be applied when instances have numeric attributes, because it is unable to distinguish any difference between numbers that are close in value from those that are far apart. For example, the real numbers 3.112, 3.113, and 12.9 would all be treated as distinct, unrelated values by the original equation. However, category utility can be adapted to deal with numeric valued attributes. Because probabilities for numeric attributes are stored as a normal distribution (a mean and a standard deviation), the innermost summation in the ordinary category utility equation can be replaced with the integral of the equation for the normal distribution

$$\sum_j P(A_i - V_{ij})^2 \longleftrightarrow \int \frac{1}{\sigma^2 2\pi} e^{-(x-\mu)^2} dx = \frac{1}{\sigma} \frac{1}{4\sqrt{\pi}}.$$

The transformed evaluation function is then

$$\frac{\sum_k^K P(C_k) \sum_i^I \frac{1}{\sigma_{ik}} - \sum_i^I \frac{1}{\sigma_{ip}}}{4K\sqrt{\pi}}$$

where K is the number of classes, I is the number of attributes, σ_{ik} is the standard deviation for attribute i in class k, σ_{ip} is the standard deviation for attribute i in the parent (i.e., where no information is present).

One problem with this transformed equation is that $\sigma = 0$ when a concept node describes a single instance, so the $1/\sigma$ is ∞ in this case. In this situation, COBWEB relies on a user-specified parameter, *acuity*, to serve as a minimum value for σ. Acuity represents the minimum detectable difference between instances.

Typically, some instance descriptions are incomplete, with values missing for one or more attributes. In this implementation of COBWEB, we adapt the category utility equations so they handle this situation by dividing the attribute summations by I, the number of attributes in the incoming instance. The revised equations are:

$$\frac{\sum_{k=1}^K P(C_k) \frac{\sum_i^I \sum_j^J P(A_i - V_{ij}|C_k)^2}{I} - \frac{\sum_i^I \sum_j^J P(A_i = V_{ij})^2}{I}}{K}$$

for discrete values, and

$$\frac{\sum_{k=1}^K P(C_k) \frac{\sum_{i=1}^I 1/\sigma_{ik}}{I} - \frac{\sum_{i=1}^I 1/\sigma_{ip}}{I}}{4K\sqrt{\pi}}$$

for continuous values. As Gennari[4] points out, mixing nominal and numeric attributes in a single instance description is an open issue in the literature on numerical taxonomy and clustering. However, Gennari[5] presents evidence that summing together terms from both forms of the equation works well in domains with mixed data.

11.4 FARLEY AND RAFTERY GAUSSIAN MIXTURE MODEL

Generally in model-based clustering, it is assumed that the data is generated by a mixture of distributions in which each component represents a different group or cluster. The model for the composite of the clusters is usually formulated either by the classification likelihood approach or by the mixture likelihood approach. Farley[6] developed efficient algorithms for hierarchical clustering with the various parametrizations of Gaussian mixture models. Ward[7] used the sum of squares, which is based on Gaussian mixtures.

Model-based clustering methods attempt to optimize the fit between the data and some mathematical model. Finite mixtures are model-based clustering approaches where probabilistic clustering algorithms model the data using a mixture of distributions. Each cluster is represented by one distribution, which governs the probabilities attribute values in the corresponding cluster. Usually the individual distributions are normal distributions, referred to as Gaussian distributions. The resultant distributions are combined using cluster weights.

Note that the probability of an instance x belonging to cluster A is

$$P(A|x) = \frac{P(x|A)P(A)}{P(x)} = \frac{\frac{1}{\sqrt{2\pi}\sigma}e^{\frac{(x-\mu)^2}{2\sigma^2}}}{P(x)}.$$

[4] Gennari, J. H. (1989). Focused concept formation. *Proceedings of the Sixth International Workshop on Machine Learning*. Ithaca, NY, Morgan Kaufmann, 379-382.

[5] Gennari, J. H. (1990). *Concept Formation: an empirical study*. Doctoral dissertation, Department of Information and Computer Science, University of California, Irvine.

[6] Farley, C. (1999). Algorithms for model-based Gaussian hierarchical clustering. *SIAM J. Sci. Comput. 20*, 270-281.

[7] Ward, J. H. (1963). Hierarchical groupings to optimize an objective function. *J. Amer. Stat. Assoc., 62*, 1159-1178.

The likelihood of an instance given the clusters is:

$$P(x|the\ distribution) = \sum_i P(x|cluster_i)P(cluster_i).$$

In model-based clustering, the data x are viewed as combing P from a mixture density $f(x) = \sum_{k=1}^{G} \tau_k f_k(x)$, where f_k is the probability density function of the observations in group k, and τ_k is the probability that an observation comes from the kth mixture component $\tau_k \in (0, 1)$ and $\sum_{k=1}^{G} \tau_k = 1$.

The following explanation of mixture methods is taken from Fraley and Raftery.[8] The article has the following URL, *http://www.jstatsoft.org/*.

Each component is usually modeled by the normal or Gaussian distribution. Component distributions are characterized by the mean μ_k and the covariance matrix Σ_k, and have the probability density function

$$\emptyset(x_i; \mu_k; \Sigma_{k'}) = \frac{\exp\left(-\frac{1}{22}(x_i - \mu_k)^T \sum_k^{-1} (x_i - \mu_k)\right)}{\sqrt{\det(2\pi\Sigma_{k'})}}.$$

For univariate data, the covariance matrix reduces to a scalar variance. The likelihood for data consisting of n observations assuming a Gaussian mixture model with G multivariate mixture components is

$$\prod_{k=1}^{n} \sum_{k=1}^{G} \tau_k \, \emptyset(x_i; \mu_k; \Sigma_{k'}).$$

For a fixed number of components G, the model parameters τ_k, μ_k, and Σ_k can be estimated using the EM algorithm initialized by hierarchical model-based clustering Fraley and Raftery.[9] Data generated by mixtures of multivariate normal densities are characterized by groups or clusters centered at the means μk, with increased density for points nearer the mean. The corresponding surfaces of constant density are ellipsoidal. Geometric features (shape, volume, orientation) of the clusters are determined by the covariances Σ_k, that may also be parametrized to impose constraints across components. There are a number of possible parameterizations of Σ_k, many of which are implemented in the R package mclust.

---

[8] Farley, C., & Raftery, A. E. (2007). Model-based methods of classification: using the mclust software in chemmometrics. *Journal of Statistical Software*, (108), 6.

[9] Farley, C., & Raaftery, A. E. (1998). "How many clusters? Which Clustering method? -Answers via Model-based cluster Analysis" *Computer Journal*, 41, 578-588.

Mixture likelihood itself or its value at given points enables density estimation. Fitted likelihood is useful for estimating and comparing data trends. Assume we know there are k clusters. Then to learn the clusters, we need to determine their parameters, or their means and standard deviations. Note that the likelihood of the training data, where the model is implemented using a neural network can be used as a performance criterion, the EM algorithm finds a local maximum of the likelihood.

In fact, the cluster probabilities can be stored as the instance weights for the neural network model. Model-based clustering allows for overlapping clusters. These model-based clusterings are referred to as *generative models*.

Assume we have generated a set of points for a Gaussian distribution. Then each cluster has been generated a Gaussian probability density function with a specified mean μ and specified variance σ^2. The probability that a point x_i belongs, or was generated by the model, to cluster X is

$P(x_i | \mu, \sigma^2) = \dfrac{1}{\sqrt{2\pi}} e^{-\frac{(x_i - \mu)^2}{2\sigma^2}}$. The likelihood that the cluster X is generated by the model is:

$$L(N(\mu, \sigma^2): X) = P(x | \mu, \sigma^2) = \prod_{i=1}^{n} \frac{1}{\sqrt{2\pi}} e^{-\frac{(x_i - \mu)^2}{2\sigma^2}} .$$

The distance between two clusters C_i and C_j then becomes:

$$\text{Distance}(C_i, C_j) = -\log \frac{P C_i \cup C_j}{P(C_i) P(C_j)}.$$

11.5 ESTIMATE THE NUMBER OF CLUSTERS

For a set of objects partitioned into m clusters C_i, for $i = 1$ to m, the quality can be measured by:

$$Q(\{C_1, ..., C_m\}) = \prod_{i=1}^{m} L(N(\mu, \sigma^2): C_i).$$

If we merge two clusters C_a *and* C_b into a cluster $C_a \cup C_b$ then the change in quality of the overall clustering is

$$Q(((\{C_1, ..., C_m\} - \{C_a, C_b\}) \cup \{C_a \cup C_b\}) - Q(\{C_1, ..., C_m\}))$$

$$= \frac{\prod_{i=1}^{m} P(C_i)P(C_a \cup C_b)}{P(C_a)P(C_b)} - \prod_{i=1}^{m} P(C_i)$$

$$= \prod_{i=1}^{m} P(C_i)\left(\frac{P(C_a \cup C_b)}{P(C_a)P(C_b)} \right) - 1).$$

The problem of determining the number of clusters is solved by choosing the "best model," or the model with maximum quality. The interested reader should study the following article by Farley and Raftery.[10]

11.6 SUMMARY

- COBWEB limitation: assuming that probability distributions on separate attributes are statistically independent of one another.

- COBWEB limitation: hierarchy tree expensive to update and store the clusters due to the probability distribution representation of clusters.

- COBWEB limitation: the classification tree is not height-balanced for skewed input data.

- COBWEB: incremental clustering algorithm, based upon probabilistic categorization trees.

- COBWEB: search for a good clustering is guided by a quality measure for partitions of data.

- COBWEB: only supports nominal attributes.

- CLASSIT: modification of COBWEB that works with nominal and numerical attributes.

- Finite Mixtures Models: allow overlapping clusters.

- Finite Mixtures Models: can be implemented as neural networks.

- COBWEB: is a probabilistic hierarchical clustering method:

   1. Use probabilistic models to measure distances between clusters.

   2. Assumption: adopt common distribution functions such as Gaussian or Bernoulli.

---

[10] Fraley, C., & Raftery, A. E. (1998). How many clusters? Which clustering method? Answers via model-based cluster analysis. *The Computer Journal*, Vol. 41, No. 8.

- Gaussian Mixture Models: are generative models, which regard the set of data objects to be clustered as a sample of the underlying data generation mechanism to be analyzed.

11.7 EXERCISES

1. Develop an algorithm for construction of a probabilistic agglomerative hierarchical clustering.

2. Explain how clusters are formed for a probabilistic hierarchical clustering?

3. Discuss the advantages and disadvantages to the application of probabilistic hierarchical clustering.

4. What are the limitations of COBWEB?

5. Explain how COBWEB determines intra-class similarity.

6. Explain how COBWEB determines inter-class similarity.

7. Develop the COBWEB algorithm.

8. Discuss the limitations to COBWEB.

9. Discuss the limitations to CLASSIT.

CHAPTER **12**

GENERAL ISSUES

In This Chapter

12.1 INTRODUCTION

Many issues are present in a cluster analysis study. Do the clusters in a resultant clustering accurately represent the data set? What is the "correct" number of clusters? Because the foundation of clustering includes the similarity or dissimilarity measure chosen, the investigator has to ask if the "correct" measure has been chosen.

Once these issues have been resolved, there are still several more issues to consider. Remembering the old GIGO phrase, garbage-in-garbage-out, leads to investigating and validating the input data set. Outliers may need to be identified and either corrected or possibly eliminated from the study. Finally, it is the matter of interpretation, for the intra-cluster and the inter-cluster meanings.

12.2 DATA CLEANSING

First the investigator needs to cleanse the data. Inclusion of unrelated attributes in a cluster analysis will negatively impact the proximity computations. These suspicious computations will then negatively hinder the clustering results, because the foundation for the neighborhoods used to capture the clusters is the proximity function. In summary, attribute selection is a major step to make for data preparation. Inappropriate attributes should be identified and eliminated. If the cluster analysis will include categorical data, then, for the sake of preserving linear independence, a minimal set of categorical attributes should be chosen. Also in the field, a common practice is to assign different weights to attributes or standardizing the attribute values to generate desirable cluster shapes. Assignment of attribute weighting can allow for placing varying degrees of importance on the attributes. In many applications, maintaining the varying degrees of attribute importance is critical. Consider the problem of capturing the sales strategies employed by the sales force for a large chain of hardware stores. Selling some product lines will have a greater impact on year-to-year profits than other product lines.

A particular method may only work on specific data types. The method in Chapters 1 through 4 are strictly numerical clustering methods and Chapter 8 deals exclusively with categorical data sets. The given data set and the required solution type for the application may necessitate converting some of the attribute data types into an appropriate type for the chosen method. In fact, when employing a fuzzy method, the attribute values will need to be transformed into a tuple format which includes the attribute value plus the associated weight.

In general, data quality is a concept represented by various intrinsic understanding or data quality dimensions, based upon viewpoint. Wandand and Wang[1] reported, utilizing a background review, that the top 5 data quality dimensions include: accuracy, reliability, timeliness, relevance, and completeness. Data cleansing is an activity used to achieve data quality.

---

[1] Wandand, Y., & Wang, R. (November, 1996). Anchoring data quality dimensions ontological foundations. *Communications of the ACM*, 39(11).

The major activities in data cleansing are analyzing data fields, removing erroneous data, filling in missing information, and enforcing business rules. Data cleansing is a continuous ongoing process. Initial validated data can become invalid over time. Dirty data can seriously corrupt a statistical or cluster analysis. The foundation for implementing a sound data cleansing is to possess understanding of the full complement of codes in the data set.

The following steps can be undertaken to achieve data quality:

1. Define the need.

2. Define the domains and application system rules.

3. Determine the current state of the data set.

4. Analyze the data set problems.

5. Count the cost and fixing.

For defining the need, several methods can be used to acquire accurate information requirements. First, asking a controlled set of questions of the users or consumers. This is a standard method of system analysis. Second, the original data set can be replaced in implementation. Another organizations data set or data sets from research or industrial studies could be used as the replacement. Third, a review of how the consumers of the information will use the data and design, from the proposed approach for the cluster analysis, should be undertaken. Finally, a prototype study should be conducted in order to allow the consumers and investigators to move from the abstract requirements into concrete examples.

When analyzing the data set problems, a review of the existing domain values and business rule enforcement will be required to ensure continued cleansed data. Methods to correct the original problems should be proposed to fix the original problem. Additionally, methods to fix existing problems should also be proposed.

Once the costs have been assessed and justified against the need for clean data, fixing data source problems can commence. When these tasks are completed, fixing the existing problems can be accomplished.

Data cleansing is more than a method for converting dirty data into clean data. It is a process to understand:

1. What must be done.

2. What impacts it will have in terms of cost, usability, and derived benefits.

Ongoing review of data collection and extraction policies, data entry training and programmatic assessment of the critical data items will help keep data from becoming the target of required cleansing.

12.3 WHICH PROXIMITY MEASURE SHOULD BE USED?

The method used often determines the choice of the proximity measure. K-means normally uses the sum of the squares of distances between points and centroids. Hierarchical clustering methods use different distances and similarity measures. However, for numeric data, the usual L_p is commonly applied where:

$$d(x, y) = \|x - y\|_p, \text{ where } \|z\|_p = \left(\prod_{j=1,d} |z_j|^p\right)^{1/p} \text{ and } \|z\| = \|z\|_2$$

in which p corresponds to a more robust estimation, $1 \le p \le \infty$. If $p = 2$, then we are using 2-dimensional Euclidean distance, often used with K-means. When $p = 1$, we are using the Manhattan distance. Many clustering studies have data points scaled to return a unit norm, in this case the proximity measure is an angle between the data points:

$$d(x, y) = \arccos\left(x^T y \bigg/ \left(\|x\| \bigg/ \|y\|\right)\right).$$

There are numerous similarity measures for categorical data. The Rand and Jaccard indices, R and J, are often used for categorical data for measuring similarity. The Jaccard index treats positive and negative values symmetrically. Because of the symmetric treatment of positive and negative values, the Jaccard index is usually the index of choice with transactional data sets.

12.4 IDENTIFYING AND CORRECTING OUTLIERS

Clustering methods do not distinguish between "normal" data points and outliers, or "abnormal" data points. One method for handling outliers is to rely on input thresholds to eliminate low-membership clusters. Assume that in Figure 12.1 the clusters illustrated meet the threshold specification.

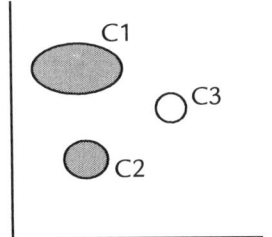

FIGURE 12.1 Identification of Clusters to Delete.

Care must be taken because outliers are identified by their local neighborhoods in some methods, especially those methods using distance-based outlier detection.

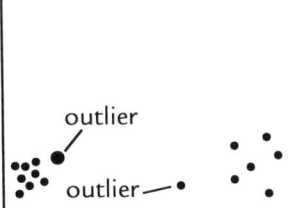

FIGURE 12.2 Distance-based Outliers are Identified by their Local Neighbors.

Because distance-based outlier detection methods are sensitive to the mean and standard deviation then clustering results can be highly suspect. Figure 12.3 illustrates such a case:

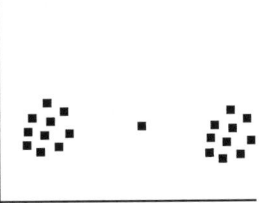

FIGURE 12.3 The Mean of the Data Set is the Outlier.

Also the need for density-based metrics becomes evident when you study Figure 12.4. For x to be an outlier for cluster C1 then all points in cluster C2 have to be declared as outliers.

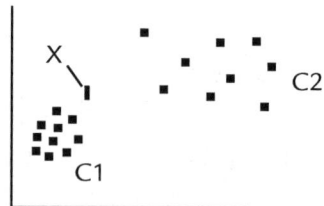

FIGURE 12.4 The Need for Density-based Outliers.

The solution to these problems is to define local outliers. The problem is that different subsets of data have different densities and could even be defined for different distributions. Points lying closer to a dense cluster can have a higher probability of being identified as an outlier than points that are a further distant from a sparse cluster. The LOF, local outlier factor, resolves this problem. An LOF specifies the degree of outlier tendency for a data point in terms of the distance to the k-nearest neighbor.

12.5 FURTHER STUDY RECOMMENDATIONS

Pattern recognition and image processing are major applications that employ cluster analysis. Many of the concepts and methods from these applications would clearly have potential for clustering in today's multimedia environment. The interested reader should find out what is meant by pixel validity and how it is assessed. Also investigate clustering methods by visualization techniques. These types of clustering methods are applicable to: images, micro-array data from Bio-informatics, and geographical data. The reader should look into clustering for image retrieval and related usage of relevant feedback. Find out what the general approach is to clustering in these applications.

Neural nets have a vast impact on clustering. Find out how these clustering methods differ from the methods presented in the text. Try to determine what are the advantages and disadvantages for using a neural net approach?

12.6 INTRODUCTION TO NEURAL NETWORKS

Perceptrons are the simplest of all neural network architecture and can serve as a first platform for developing an understanding of neural networks. A perceptron network consists of an input layer of nodes fully connected to an output layer of nodes as illustrated in Figure 12.5.

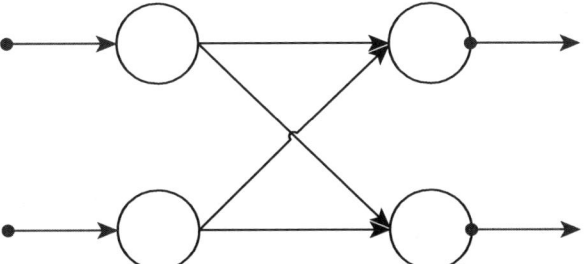

FIGURE 12.5 Two-feature Two-layer Perceptron.

The number of nodes in the output layer corresponds to the number of classes. The number of nodes in the input layer is represented by the number of input features.

Weights are assigned to each connection between the input and output layers as in Figure 12.6. Let x_i represent the ith input to the ith input node and w_{ij} is the jth weight associated with the ith input. Then net_i is the ith net input to the perceptron as illustrated in Figure 12.6.

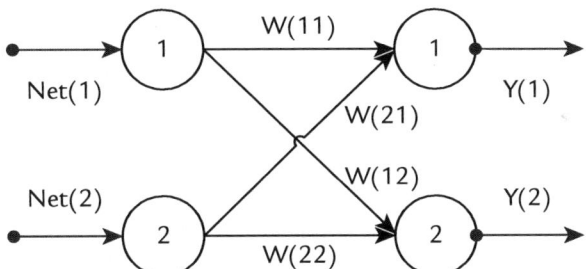

FIGURE 12.6 Weight Assignments for the Perceptron in Figure 12.5.

$$net_i = \sum_{j=1}^{2} x_i w_{ij}$$

An activation function is applied to the inputs, usually the sigmoid function:

$$Y_i = \frac{1}{1 + e^{-net_i}}.$$

Once the perceptron has been constructed, then it has to be trained. Activating inputs enables the comparison of the output to the desired, or target, output. Let Y_i be the ith output and Y_i' be the ith target output.

Then the weights can be updated by the differences in the output and target outputs, or

$$w_i(k+1) = \alpha(Y_i' - Y_i)x_i$$

where k is the iteration number and α is the learning rate. This iterative process, like the Kelley-Salisbury Method for regression discussed in Chapter 5, is repeated until there is no significant change between the output vector and the target output vector or a maximum number of iterations has been reached.

Consider training a two-input/one-output perceptron to obtain an output of –1.

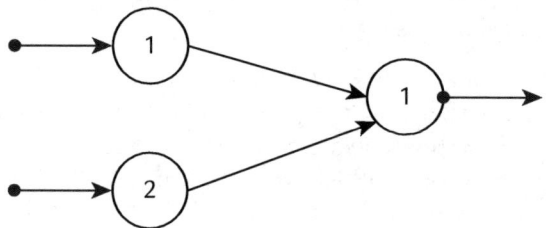

FIGURE 12.7 Two-Input/One-Output Perceptron.

Begin the training by randomly assigning weights. Note that

$$w_i(k+1) = w_i(k) + \Delta w_i(k)$$

and

$$\Delta w_i(k) = (Y_i' - Y_i)x_i(k)$$

where k is the iteration number.

For example, suppose that $Y' = 1$, $Y = 0$, $w_1 = 0.5$, $w_2 = 0.3$, $x_1 = 2$, $x_2 = 1$, $\alpha = 1$

then, on the next iteration:

$$w_1(k + 1) = 0.5 + (0 - 1)(2) = -1.5$$
$$w_2(k + 1) = 0.3 + (0 - 1)(2) = -0.7.$$

If we input the new weights, then the perceptron will output 0 instead.

$$net_{k+1} = w_1(k + 1)x_1 + w_1(k + 1)x_1$$
$$net_{k+1} = (-1.5) * 2 + (-0.7) * 1 = -3.7$$

$$Y' = \frac{1}{1 + e^{-(-3.7)}} = 0.024 \text{ which implies } y' = 0.$$

Therefore, the iteration process must continue at this point.

Neural networks are literally used for making classifications. Addition of the learning rate will speed up the learning process. In fact, neural networks extend the classification problems that can be solved by the traditional methods discussed in this textbook.

For example, consider building a perceptron for an exclusive or operation. The inputs and outputs for this operation are:

Inputs	Outputs
(0,0)	0
(0,1)	1
(1,0)	1
(1,1)	0

TABLE 12.1 Exclusive or Truth Table.

First, consider the fact that if a linear discriminant function exists that can separate the classes without error, then it can be shown that the training procedure for the associated discriminant is guaranteed to find that line or plane. However, such a discriminant function does not exist for the exclusive or operation, this problem is not linearly separable. Neural network architectural design resolves the linearly nonseparable problem by utilizing a multilayered architecture.

Rumelhard, McClelland, and the PDP Group[2] proposed a *back-propagation network* for resolving classification in nonlinear separable data. This model consists of a minimum of three layers: an input layer, one or more hidden layers, and an output layer. Nodes are fully connected between layers and use the sigmoid activation function. Training uses the gradient descent method to minimize the sum of the squared error.

Output is obtained through using the *feed-forward approach*. The inputs for each layer, except for the input layer, are found using the previously

---

[2] Rumelhart, D. E., McClelland, J. L., & the PDP Group. (1986). Learning internal representation by error propagation. *Parallel Distributed Processing Models Explorations in Microstructure of Cognitron,* vol II. Cambridge, MA: MIT Press.

discussed weighted-sum method. Components for the output vector are found by the previously stated sigmoid activation function. The change in weights is obtained using a generalization of the delta learning rule. For this network, the error at the output layer is propagated back through the network to compute the change in weights. Figure 12.11 illustrates a back-propagation neural network.

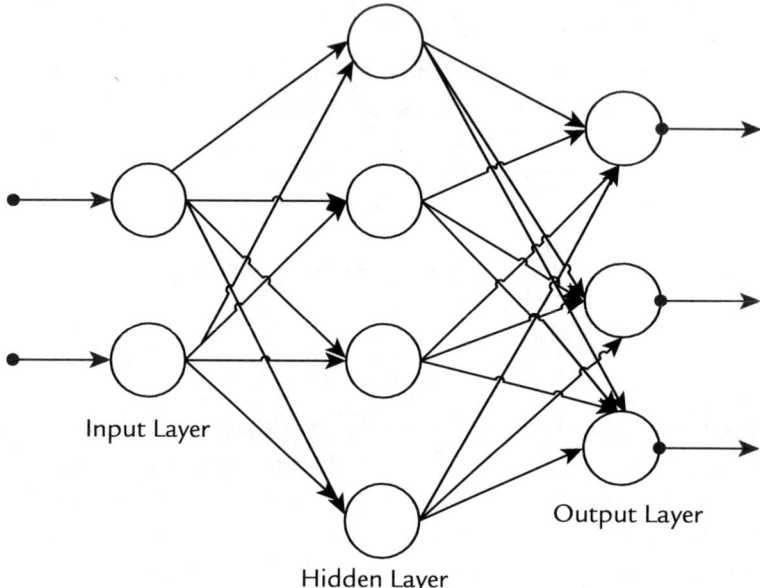

Input Layer

Hidden Layer

Output Layer

FIGURE 12.8 Two Feature, Three Class Back-Propagation Neural Network.

The objective is to minimize the error:

$$\text{Error} = (1/2)\sum_i (Y_i' - Y_i)^2.$$

The direction on the error surface which most rapidly reduces the error needs to be derived, or we must find the slope of the error function. Finding the slope, by taking the derivative, is referred to as the *gradient descent method.*

$$\Delta w_i = -\frac{\partial Error}{\partial w_i}, \text{ where } Y_i = f(x_i, w_i)$$

and

$$\frac{\partial Y_i}{\partial w_j} = x_j f'(x_j w_j), \text{ where } f'(x_j w_j) \text{ is the activation function.}$$

Using the chain rule for differentiation yields

$$\frac{\partial Error}{\partial w_j} = \frac{\partial Error}{\partial w_j}\frac{\partial Y_j}{\partial w_j}.$$

However, $\Delta w_j = -c\left(-\left(Y_j' - Y_j\right)\right)x_i f'(x_i, w_i),$

then $$\frac{\partial Error}{\partial Y_j} = \frac{\partial\left(\frac{1}{2}\right)\left(\sum_i (Y_i' - Y_i)^2\right)}{\partial Y_j'} = (Y_j' - Y_j).$$

The sigmoid function is a good choice because the activation function needs to be differentiable.

Let L_1 represent the input layer, L_2 represent the hidden layer, and L_3 represent the output layer. $\overrightarrow{X}$ is the input vector, $\overrightarrow{T}$ is the target output vector, and o_i is the ith component of the actual output vector. Let α be the learning rate between L_2 and L_3, and β be the learning rate between L_1 and L_2.

The change in weights between L_2 and L_3 is:

$$\delta_i = (t_i - o_i)o_i(1 - o_i)$$

$$\Delta w_{ij} = -\alpha\delta_i o_j$$

Because there is no target vector with which to compare the hidden layers, the following equation is used to change the weights between layers L_1 and L_2. Note that o_{ij} represents the output of component j in the hidden layer and o_i represents the ith component in the output vector.

$$\Delta w_{ij} = -\beta o_{ij} o_i (1 - o_i)\sum_{k=1}^{m}\delta_k w_{ik}$$

After the weights are updated, the error is calculated using the following equation:

$$Error = \frac{1}{2}\sum_{i=1}^{l}(o_i - t_i)^2.$$

This process is repeated until the error is less than some minimum-error tolerance.

Often neural networks are viewed as "black boxes" that successfully classify data, but without any explanation of how the network reached the decisions. The extraction of fuzzy if-then rules from *fuzzy neural networks* provide potential resolution to equipping neural networks with the needed explanations of how the network decisions were made in a data classification. The interested reader should study Ishibuchi, Nii, and Turksen.[3]

Example 12.1

Consider the problem of estimating the cost of a house based upon the price of comparable houses. Input data would include: location, square footage, number of bedrooms, lot size in acreage, number of bathrooms, etc. One output is desired, the sales price. Training data includes features of recently sold houses and the related sales prices.

Input data consists of discrete data (number of bedrooms, number of bathrooms), categorical data (location), and continuous data (square footage, lot size in acreage). Typical input to output is illustrated in Figure 12.12.

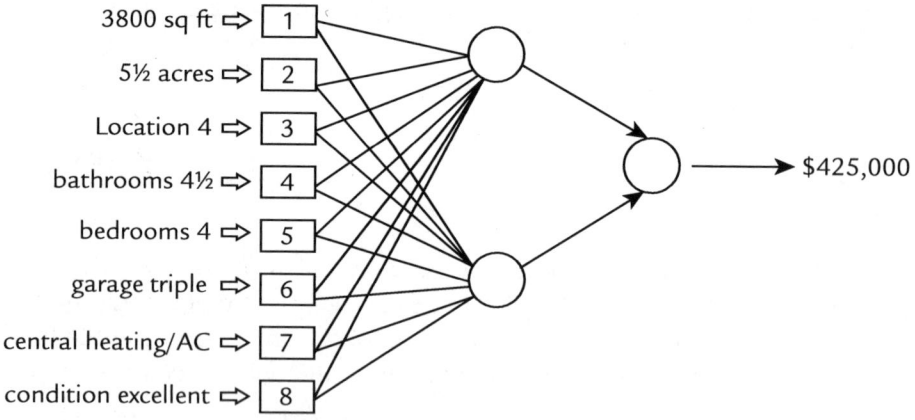

FIGURE 12.9 Intuitive Neural Network Design.

---

[3] Ishibuchi, H., Nii, M., & Turksen, J. B. (1998). Bidirectional Bridge between Neural Networks and Linguistic Knowledge: Linguistic Rule Extraction and Learning from Linguistic Rules. *Proceedings of the IEEE Conference on Fuzzy Systems FUZZ.IEEE '98*, Anchorage, AK., 1112-1117.

After training, the neural network can be validated by using a set of sales examples never seen before by the network. This can also be accomplished by dividing the accessible data before training into a training set and a validation set.

Example 12.2

Consider the iris plant classification problem and data set available from the University of California at Irvine Machine Learning Laboratory. The input data set consists of four variables (sepal length, sepal width, petal length, and pedal width) for three classes of iris plants (setosa, versicolor, and virginica). An intuitive neural network for this classification is illustrated in Figure 12.10.

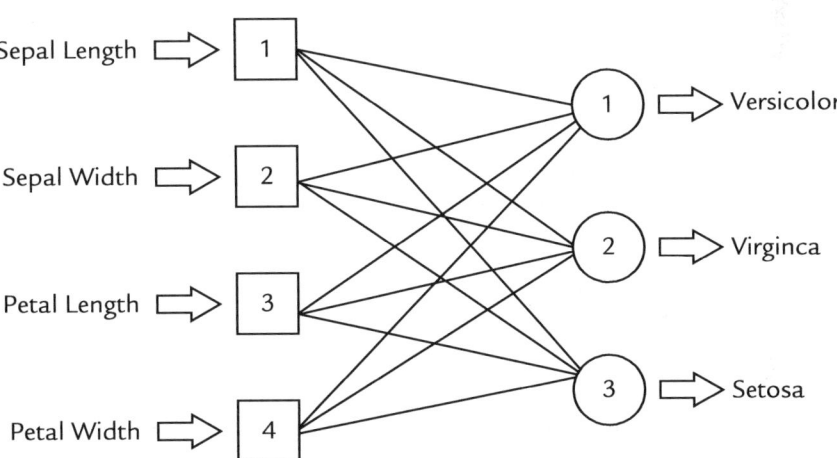

FIGURE 12.10 Intuitive Neural Network for the Iris Classification.

From the 150 cases in the data set, the investigator should randomly divide the cases into a training and a validating set. This network is trained to divide input vectors into three classes [(1,0,0), (0,1,0), and (0,0,1)]. This is a *competitive neural network* and does not possess the ability to compute the sum of squared errors criterion because the desired outputs are unknown. Therefore, simply stop when no noticeable changes occur in the weight vectors, which is a sign that the network has converged. Clearly neural networks enable one to identify clusters in input data. In fact, a competitive neural network can classify input patterns that are not linearly separable. The method for a competitive neural network learning an instance of an input is divided into four steps:

Step 1: Determine the output for the new input instance.

Step 2: Determine the difference between the output and the target output for each output node. These differences are linear error, therefore they are proportional to partial derivatives and then we can process the gradient descent.

Step 3: Determine, for all input and hidden nodes, $w_k * error_k$, where $error_k$ is the difference found in Step 2 and w_k is the weight of edge that connects to the output node. This step propagates the error backwards, and explains why the learning method is called back-propagation.

Step 4: The gradient descent is performed using the formula:

$$w_{ij} = w_{ij} + r * error_j * a_j * x_j + o_i,$$

Where r is the learning rate, $error_j$ is the difference between the output and target output for output node j, and x_j, as well as o_j, are determined as in Step 1. a_j is the partial derivative of output node j's activation function. The method terminates when the training instances have been run, at this point learning is complete.

The human brain maintains centers for speech, vision, hearing, and motor functions. These centers are located in areas next to each other. Kohonen[4] developed neural networks where ordered feature maps develop naturally. Output units in the neural network located physically next to each other are made to respond to classes of input vectors that are also next to each other. Kohonen neural networks are extensions of competitive learning networks where input units are ordered, often in a two-dimensional grid. The ordering allows the user to determine which output nodes are neighbors. This topology-preserving map is referred to as a *self-organizing map, SOM*. As input is submitted to the SOM network the following learning rule applies to the winning output node i:

$$\vec{w}_0(t+1) = \vec{w}_0(t) + g(0, k)(\vec{x}(t) - \vec{w}_0(t)), \text{ for every } 0 \in S.$$

Where t is time, $\vec{x}(t)$ is an input vector presented to the network, S is the set of output nodes, and $g(0,k)$ is a decreasing function of the grid-distance between nodes 0 and k where $g(k,k)$. $g(0,k) = e^{-(-(0-k)2)}$ if $g(0,k)$ is the Gaussian function.

---

[4] Kohonen, T. (1988). *Self-Organizing and Associative Memory*, 3rd ed. Springer Verlag, New York.

During each training period, each unit with a positive activity within the neighborhood of the winning output node participates in the learning process. The interested reader should consult *http://wwww.cs. hmc.edu/~kpang/nn/som.html* and *http://www.rocksolidimagess.com/pdf/ Kohonen.pdf*.

12.7 INTERPRETATION OF THE RESULTS

Attribute selection and each tuple's feature or attribute values will impact the interpretation of the resultant clustering. Therefore, the results of the cluster analysis need to be validated or at least evaluated. A panel of experts in the domain application and in cluster analysis could assess the face validity of the resultant clustering. Kandogan[5] discusses cluster visualization which can be used as an alternative for cluster validation. One of the main observations formed after reviewing research studies and journal articles on clustering is that the interpretability depends on the method. For instance, K-means generates clusters that are dense neighborhoods with the centroids as the cluster centers. Kandogan[1] uses cluster visualization to validate.

The Rand statistic, or the Jaccard statistic, which allows for comparison of two distinct clusterings, can be used for validation, especially in Monte Carlo studies. Other validation methods are based on conditional entropy by Cover and Thomas[6] and using the F-measure by Larsen and Aone.[7]

12.8 CLUSTERING "CORRECTNESS"?

Consider the following graphs for the same data set in Figure 12.11.

Sometimes more than one clustering is representative of the data set.

Not only is the question concerning the number of clusters important but another major question is whether or not distinct clusters are equally representative of the data set, especially where densities vary.

---

[5] Kandogan, E. (2001). Visualizing multi-dimensional clusters, trends, and outliers using star coordinates, In *Proceedings of the 7th ACM SIGKDD*, San Francisco, CA, 299-304.

[6] Cover, T. M., and Thomas, J. A. (1999). *Elements of Information Theory*, John Wiley & Sons, New York, N. Y.

[7] Larsen, B. and Aone, C. (1999). Fast and efficient text mining using linear-time document clustering, In *Proceedings of the 5th ACM SIGKDD*, San Diego, CA, 16-22.

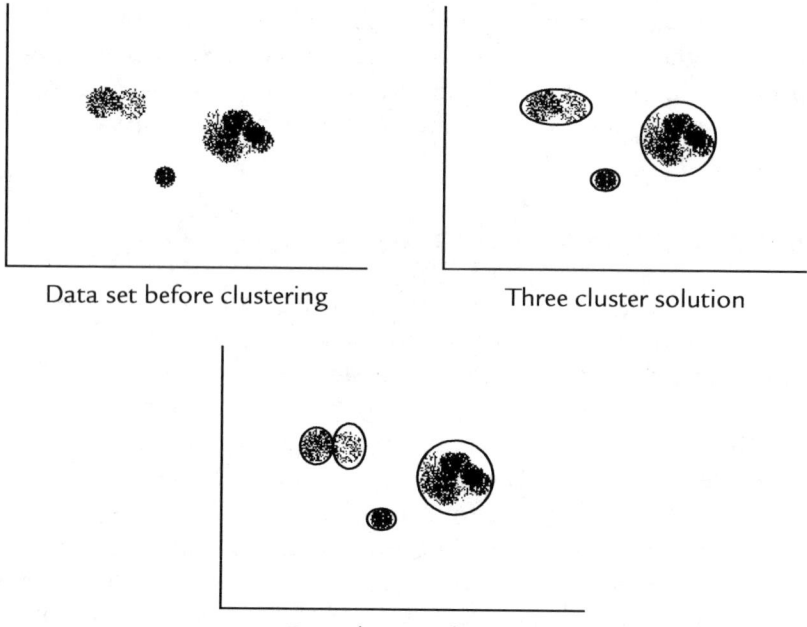

Data set before clustering   Three cluster solution

Four cluster solution

FIGURE 12.11 Two Potentially Correct Clusterings.

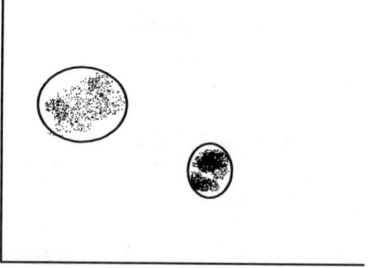

FIGURE 12.12 Clusters with Unequal Densities.

During a clustering study, multiple runs using a specific method generates a sequence of clustering results. Successive runs produce clusters that tend to contain clusters closer together. Many times these results are due to the method's underlying function. The objective function for the K-means method is a monotone decreasing function and, therefore, magnifies the sequence of clustering results. Which cluster does x in Figure 12.13 best qualify for membership?

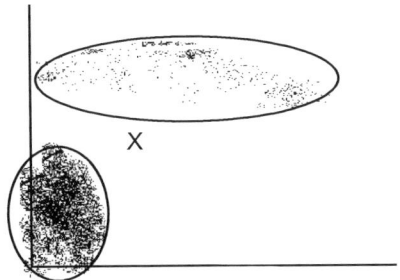

FIGURE 12.13 Which Membership for Data Point *x*?

Because the K-means method is based upon ANOVA, Milligan and Cooper[8] used the F statistic to find an optimal number of clusters in a data set. Another useful observation is that a choice for establishing a clustering coefficient would be the distance between two centroids normalized by the corresponding cluster's radii, which measures cluster spread, and averaged by cluster weights. Kaufman and Rousseeuw[9] define a Silhouette coefficient which finds the average distance to the best fitting cluster compared to the average distance between a data point $x \in C$ and other points of C for determining cluster system appropriateness. Cohesion measures how closely related objects are in a cluster, while separation measures how distinct or well-separated a cluster is from other clusters.

Definition: *Cohesion a(x)*: average distance of x to all other vectors in the same cluster.

Definition: *Separation b(x)*: average distance of x to the vectors in other clusters.

To find the minimum among the clusters:

Definition: *Silhouette s(x)*:

$$s(x) = \frac{b(x) - a(x)}{\max\{a(x), b(x)\}}$$

A value of +1 indicates a perfect clustering choice and a value below 0 indicates a bad clustering choice.

---

[8] Milligan, G., & Cooper, M. (1985). An examination of procedures for determining the number of clusters in a data set. *Psychometrika*, 50, 159-179.

[9] Kaufman, L., & Rousseeuw, P. (1990). *Finding Groups in Data: An Introduction to Cluster Analysis*. New York: John Wiley and Sons.

Definition: *Silhouette coefficient (SC)*:

$$SC = \frac{1}{N}\sum_{i=1}^{N} s(x).$$

Bezdek[10] addresses the concept of separation using fuzzy assignments. For a particular cluster C, different weights $w(x,C)$ are used to define $C(x) = \text{minarg}_C \, w(x,C)$. Then a partition coefficient is defined to equal the sum of squares of the weights:

$$W = \frac{1}{N}\prod_{x \in X}(C(X))^2.$$

The best choice for k, the number of clusters, can be determined by plotting for each of the measures, discussed above, as a function of k.

12.9 TOPICAL RESEARCH EXERCISES

The following exercises are for the reader to perform topical background searches. You are to review the most cited topics in each problem area, identify and illustrate the application of the techniques or methods in the problem area, and explain the motivation for their development. Perform a topical background search on one or more of the following:

1. Cluster tendency rather than cluster validation.

2. Subspace outliers.

3. MINDS, the Minnesota Intrusion Detection System.

4. Sequence outliers.

5. Outliers for streaming data.

6. The "Kernel Trick" in Hilbert space.

7. Clustering in nonlinearly separable space.

8. Neural nets and clustering.

9. Projected clustering.

10. Apriory for mining frequent sets.

---

[10] Bezdek, D. (1981). *Pattern Recognition with Fuzzy Objective Function Algorithms.* New York: Plenum Press.

INDEX

21